T0145338

Thermoelectric Thin Films

Paolo Mele • Dario Narducci • Michihiro Ohta
Kanishka Biswas • Juan Morante • Shrikant Saini
Tamio Endo
Editors

Thermoelectric Thin Films

Materials and Devices

 Springer

Editors
Paolo Mele
SIT Research Laboratories, Omiya campus
Shibaura Institute of Technology (Tokyo)
Tokyo, Japan

Michihiro Ohta
AIST, Tsukuba, Japan

Juan Morante
IREC, Sant Adrià de Besòs, Spain

Tamio Endo
Japan Advanced Chemicals
Atsugi, Japan

Dario Narducci
University of Milano-Bicocca
Milano, Italy

Kanishka Biswas
JNCASR, Bangalore, India

Shrikant Saini
Department of Mechanical and Control
Engineering
Kyushu Institute of Technology
Kitakyushu, Japan

ISBN 978-3-030-20045-9   ISBN 978-3-030-20043-5   (eBook)
https://doi.org/10.1007/978-3-030-20043-5

© Springer Nature Switzerland AG 2019
This work is subject to copyright. All rights are reserved by the Publisher, whether the whole or part of the material is concerned, specifically the rights of translation, reprinting, reuse of illustrations, recitation, broadcasting, reproduction on microfilms or in any other physical way, and transmission or information storage and retrieval, electronic adaptation, computer software, or by similar or dissimilar methodology now known or hereafter developed.
The use of general descriptive names, registered names, trademarks, service marks, etc. in this publication does not imply, even in the absence of a specific statement, that such names are exempt from the relevant protective laws and regulations and therefore free for general use.
The publisher, the authors, and the editors are safe to assume that the advice and information in this book are believed to be true and accurate at the date of publication. Neither the publisher nor the authors or the editors give a warranty, express or implied, with respect to the material contained herein or for any errors or omissions that may have been made. The publisher remains neutral with regard to jurisdictional claims in published maps and institutional affiliations.

This Springer imprint is published by the registered company Springer Nature Switzerland AG.
The registered company address is: Gewerbestrasse 11, 6330 Cham, Switzerland

Preface

This book originates from the Symposium A-5 "Thermoelectric materials for sustainable development" organized in the framework of the conference IUMRS-ICAM 2017 (International Union of Materials Research Societies-International Conference on Advanced Materials) from August 27 to September 1, 2017, at Yoshida Campus of Kyoto University, in Kyoto, Japan (http://www.iumrs-icam2017.org/).

The symposium A-5 (http://www.iumrs-icam2017.org/program/forums/a-5. html) was focused on state of the art of thermoelectrics, from materials science to applications. A variety of topics were covered: metals, alloys, oxides, sulfides, selenides, and tellurides as thermoelectric materials, in the form of thin films, bulk, and single crystals, considering synthesis and characterization, device development, theory, and applications.

Symposium A-5 was highly successful, extending for 5 days (Monday to Friday), with 19 sessions in parallel (it was one of the only three symposia in IUMRS-ICAM organized with parallel sessions).

Symposium A-5 counted 183 talks (including 10 keynotes and 30 invited talks)—which made it the biggest symposium in IUMRS-ICAM 2017—and 61 posters. High level of presentations and lively discussion contributed to the quality of the symposium.

Symposium A-5 was enriched with the 2nd AAT Summer School on Thermoelectrics, organized by AAT (Asian Association of Thermoelectrics) as pre-event of the conference on August 27 (counting about 80 attendees), and satellite workshop on "Advanced Materials and Principles to Develop Viable Thermoelectrics & Effective Thermal Management" held at NIMS, Tsukuba, on September 2 (counting about 40 attendees).

Social part was also included, with symposium dinner in a famous hotel on Kamo River, the pulsing heart of the historical Kyoto City.

Symposium A-5 was one of the "Bilateral MRS-J/E-MRS" symposia at IUMRS-ICAM 2017 and counted also as the second edition of the AAT conference. Financial support from MRS-J and endorsement by AAT, THO (Team Harmonized Oxides, Japan), and AIT (Italian Thermoelectric Society) are gratefully acknowledged.

Many outstanding oral and invited presentations were given during the symposium. The symposium organizers were inspired by them to disclose such excellent papers to the widest scientific community. One of the hot topics of the symposium—the thermoelectric thin films—was selected as subject of a focused research collection. This is a reason why we invited our distinguished colleagues to share their results and we publish this book entitled *Thermoelectric Thin Films: Materials and Devices*.

Tokyo, Japan	Paolo Mele
Milano, Italy	Dario Narducci
Tsukuba, Japan	Michihiro Ohta
Bangalore, India	Kanishka Biswas
Sant Adrià de Besòs, Spain	Juan Morante
Kitakyushu, Japan	Shrikant Saini
Atsugi, Japan	Tamio Endo
March 11, 2019	

Introduction

Thermoelectricity is a well-known phenomenon enabling the conversion of heat into electric energy without moving parts. Its exploitation has been widely considered to contribute to the increasing need for energy along with the concerns about the environmental impact of traditional fossil energy sources.

Macroscale heat harvesting (macroharvesting) by thermoelectricity is meeting an increasing interest, from automotive to waste heat recovery from industrial plants. A much less considered application of thermoelectricity is microharvesting, namely, once again the conversion of heat into electricity but at the milliwatts scale. Microharvesting is sensibly one of the possible routes to provide full autonomy to sensing nodes to be deployed within wireless sensor networks, a key component of the so-called Internet of Things. To this aim, the availability of efficient thin-film thermoelectrics is of paramount relevance, also in view of the possibility of their integration within microelectronic devices. Such a request has contributed over the last decade to a surge of interest toward thermoelectric materials that may be effectively obtained as thin films.

This research effort has naturally met with a host of preparation and processing techniques making use of nanotechnology easily implantable in thin films. As an example, the insertion of nanodefects as phonon scatterers is regarded as a promising way to improve the performance of thermoelectric oxides. Compared to bulk materials, in thin films, it is much easier to design nanostructures and to tailor defect size and distribution through nanoengineering approach using a variety of techniques (chemical and physical depositions). Thus, research on nanoengineered thin films and device-based is one of the emerging fields in thermoelectric community.

This book aims to collect the state-of-the-art techniques and methodologies involved in thermoelectric thin-film growth, characterization, and device processing. His book involves widespread contributions on several categories of thermoelectric thin films: oxides, chalcogenides, iodates, nitrides, and polymers. It is a unique book which will cover a wide spectrum of topics related to thermoelectric thin films: from organic chemistry to devices, from physical chemistry to applied physics, and from synthesis to device implementation.

As the first available book solely dedicated to thermoelectric thin films, this will be invaluable to the experts to consolidate their knowledge and provide insight and inspiration to beginners wishing to learn about thermoelectric thin films.

Contents

Contributors

Björn Alling Department of Physics, Chemistry, and Biology (IFM), Linköping University, Linköping, Sweden

Max-Planck-Institut für Eisenforschung GmbH, Düsseldorf, Germany

Ananya Banik New Chemistry Unit, Jawaharlal Nehru Centre for Advanced Scientific Research (JNCASR), Bangalore, India

Kanishka Biswas New Chemistry Unit, Jawaharlal Nehru Centre for Advanced Scientific Research (JNCASR), Bangalore, India

School of Advanced Materials, Jawaharlal Nehru Centre for Advanced Scientific Research (JNCASR), Bangalore, India

International Centre for Materials Science, Jawaharlal Nehru Centre for Advanced Scientific Research (JNCASR), Bangalore, India

Lidong Chen State Key Laboratory of High Performance Ceramics and Superfine Microstructures, Shanghai Institute of Ceramics, Chinese Academy of Sciences, Shanghai, China

Per Eklund Department of Physics, Chemistry, and Biology (IFM), Linköping University, Linköping, Sweden

Seungwoo Han Department of Nano Applied Mechanics, Korea Institute of Machinery and Materials, Daejeon, South Korea

Sit Kerdsongpanya Department of Physics, Chemistry, and Biology (IFM), Linköping University, Linköping, Sweden

Western Digital Corporation, Magnetic Heads Operations, Materials Science Labs, Ayutthaya, Thailand

Edoardo Magnone Department of Chemistry and Biochemical Engineering, Dongguk University, Seoul, Republic of Korea

Paolo Mele SIT Research Laboratories, Omiya campus, Shibaura Institute of Technology (Tokyo), Tokyo, Japan

Hiromichi Ohta Research Institute for Electronic Science, Hokkaido University, Sapporo, Japan

Suresh Perumal Department of Physics and Nanotechnology, SRM Institute of Science and Technology, Chennai, Tamil Nadu, India

Matukumilli V. D. Prasad Theoretical Sciences Unit, School of Advanced Materials (SAMat), Jawaharlal Nehru Centre for Advanced Scientific Research (JNCASR), Bangalore, India

Shrikant Saini Department of Mechanical and Control Engineering, Kyushu Institute of Technology, Kitakyushu, Japan

Wei Shi State Key Laboratory of High Performance Ceramics and Superfine Microstructures, Shanghai Institute of Ceramics, Chinese Academy of Sciences, Shanghai, China

Masayuki Takashiri Department of Materials Science, Tokai University, Hiratsuka, Kanagawa, Japan

Yohann Thimont CIRIMAT, Université de Toulouse, CNRS, Université Toulouse 3 Paul Sabatier, Toulouse Cedex 9, France

Umesh V. Waghmare Theoretical Sciences Unit, School of Advanced Materials (SAMat), Jawaharlal Nehru Centre for Advanced Scientific Research (JNCASR), Bangalore, India

Qin Yao State Key Laboratory of High Performance Ceramics and Superfine Microstructures, Shanghai Institute of Ceramics, Chinese Academy of Sciences, Shanghai, China

About the Editors

Paolo Mele is currently professor at SIT Research Laboratories, Omiya Campus, Shibaura Institute of Technology, Tokyo, Japan. He obtained a master's degree in Chemistry and PhD in Chemical Sciences at Genova University (Italy). In 2003, he moved to ISTEC-SRL in Tokyo to study melt-textured ceramic superconductors. Then, he worked as postdoc at Kyoto University (JSPS Fellowship) from 2004 to 2007, at Kyushu Institute of Technology (JST Fellowship) from 2007 to 2011, at Hiroshima University (as lecturer) from 2011 to 2014, and at Muroran Institute of Technology (as associate professor) from 2015 to 2018 before reaching his current position. His research interests include materials for energy and sustainable development (superconductors and thermoelectrics); fabrication and characterization of thin films of oxides, ceramics, and metals; study of the effect of nanostructuration on the physical properties; thermal transport; and vortex matter. He is the author of more than 100 papers in international scientific journals and 4 book chapters and has 2 patents and has contributed to hundreds of communications at international conferences. He edited three books for Springer.

Dario Narducci obtained his PhD in Chemistry at the University of Milan. From 1988 to 1990, he was postdoctoral fellow at IBM T.J. Watson Research Center. In 1990, he re-joined the University of Milan as an assistant professor, moving in 1997 to the University of Milano Bicocca, where he became associate professor of Physical Chemistry in 2000. His research interests have focused on the physical chemistry of silicon and on the transport properties of disordered materials. Since 2008, Narducci has developed an intense research activity on thermoelectricity for microharvesting. Since 2010, he is the chief technical officer of a start-up developing silicon-based thermoelectric generators. He is currently involved in the ERC NanoThermMA project and is coordinating a Marie Skłodowska-Curie Global Fellowship in collaboration with MIT to develop hybrid photovoltaic-thermoelectric generators. He is currently the president of the Italian Thermoelectric Society and served as the treasurer of the European Thermoelectric Society. Author of more than 100 publications, Narducci also wrote books on nanotechnology and on hybrid thermoelectric-photovoltaic solar harvesters and filed 15 patents as well.

Michihiro Ohta received his PhD from the Kyushu Institute of Technology in 2002. He was a postdoctoral fellow at the National Institute for Materials Science (NIMS) and the Muroran Institute of Technology before joining the National Institute of Advanced Industrial Science and Technology (AIST) in 2006. He has been a senior researcher at AIST since 2013. He was a visiting scholar at Argonne National Laboratory and Northwestern University from 2011 to 2012. He is a board member of the Thermoelectrics Society of Japan. Ohta is also a technical advisor at the start-up company, Mottainai Energy, founded in 2016. His research focuses on the exploration of sulfides and nanostructured materials for thermoelectrics.

Kanishka Biswas obtained his MS and PhD degree from the Solid State Structural Chemistry Unit, Indian Institute of Science (2009), under supervision of Prof. C. N. R. Rao and did postdoctoral research with Prof. Mercouri G. Kanatzidis at the Department of Chemistry, Northwestern University (2009–2012). He is an associate professor in the New Chemistry Unit, Jawaharlal Nehru Centre for Advanced Scientific Research (JNCASR), Bangalore. He is pursuing research in solid-state chemistry, thermoelectrics, topological materials, 2D materials, perovskite halides, and water purification. He has published 95 research papers, 1 book, and 5 book chapters. He is a young affiliate of the World Academy of Sciences (TWAS) and an associate of the Indian Academy of Sciences (IASc), Bangalore, India. He is also recipient of Young Scientist Medal (2016) from the Indian National Science Academy (INSA), Delhi, India, and Young Scientist Platinum Jubilee Award (2015) from the National Academy of Sciences (NASI), Allahabad, India, IUMRS-MRS Singapore Young Researcher Merit Awards in 2016, and the Materials Research Society of India Medal in 2017. He has also received Young Scientist Wiley Award from IUMRS 2017 in Kyoto, Japan. He is selected as emerging investigator by the *Journal of Materials Chemistry C* (2017) and *Chemical Communications* (2018), Royal Society of Chemistry (RSC).

Juan Morante is, since 1985, full professor of the Faculty of Physics of the University of Barcelona. Since 2009, he has been the director of the advanced materials for energy area of the Energy Research Institute of Catalonia, IREC, and since the end of 2015, he has been appointed as director of this institute. Previously, he has been vice dean and dean of the Faculty of Physics of the University of Barcelona, director of the Department of Electronics of this university, head of studies in Electronic Engineering, and co-coordinator of the interuniversity master between the University of Barcelona and the Polytechnic University of Catalonia of the master on Engineering in Energy. His activities have been centered in electronic materials and devices, the assessment of their related technologies, and production processes, especially emphasizing materials technology transfer. He was involved on sensors, actuators, and microsystems, especially on chemical sensors. Currently, he is focused in the mechanisms of energy transfer in solid interfaces involving electrons, photons, and phonons as well as chemicals. Likewise, he is specialized in the development of renewable energy devices and systems for applications in the field of energy and environment based on nanostructures and their functionalization.

He has coauthored more than 600 publications with more than 17,500 citations ($h > 68$) according to the Scopus database and 21 patents, has directed 40 doctoral theses, has participated/coordinated numerous projects in different international and industrial programs (>50), organized various international technological scientific conferences in the field of sensors/microsystems and "nano-energy," and has been distinguished with the medal Narcís Monturiol of the Generalitat de Catalunya. He has also served as vice president of the European Materials Research Society and is the editor-in-chief of the *Journal of Physics D: Applied Physics*.

Shrikant Saini is a researcher at the Department of Mechanical and Control Engineering at Kyushu Institute of Technology, Kitakyushu, Japan. He obtained his PhD in Mechanical Engineering from the Jeju National University, South Korea, in 2011. He has worked in various institutes as a researcher such as the Institute of Technology (IIT) Kanpur, India; Jeju National University, South Korea; Hiroshima University, Japan; University of Utah, USA; and Muroran Institute of Technology, Japan. Dr. Saini has to date published more than 30 peer-reviewed research articles in international journals and 2 US patents (applied). His current research interest is energy harvesting/conversion materials especially thermoelectric and superconducting materials.

Tamio Endo holds PhD (Kyoto University, Japan) and MS (Gifu University, Japan) degrees. He is emeritus professor at Mie University (Japan), Gifu University special researcher (Japan), honorary professor of Southwest Jiaotong University (China), and visiting researcher at the University of California, San Diego, 1995 (USA). He is currently special adviser of Japan Advanced Chemicals in Atsugi, Japan. His research interests include oxide thin films, heterostructures, plasma effects, and bonding of polymer films. He has been part of many international academic projects such as Japan-India Cooperative Science Program. He has been organizer and plenary speaker of many international conferences and has given many foreign university guest talks and a representative of Team Harmonized Oxides.

Chapter 1
Thin Films of Bismuth-Telluride-Based Alloys

Masayuki Takashiri

1.1 Introduction

Thin films of bismuth telluride (Bi_2Te_3), antimony telluride (Sb_2Te_3), and bismuth selenide (Bi_2Se_3) are expected to be useful as miniaturized thermoelectric power generators in energy-constrained embedded systems owing to their excellent electrical and thermal properties near 300 K [1–4]. Their alloys have rhombohedral tetradymite-type crystal structures with the space group $D_{3d}^5 \left(R\bar{3}m \right)$ and hexagonal unit cells [5]. As an example, the crystal structure of Bi_2Te_3 is described in Fig. 1.1. The unit cell is composed of five covalently bonded monatomic sheets along the c-axis in the sequence $-Te^{(1)}-Bi-Te^{(2)}-Bi-Te^{(1)}-$. The superscripts (1) and (2) denote two different chemical states of the anions. The bonds between $Te^{(1)}$ and Bi include both covalent and ionic bonds, while $Te^{(2)}$ and Bi are purely bonded by a covalent bond. A very weak van der Waals attraction exists between neighboring $Te^{(1)}$ layers. The crystal structures of Sb_2Te_3 and Bi_2Se_3 are equal to that of Bi_2Te_3, but their lattice parameters are different, as presented in Table 1.1 [6]. The lattice parameter along the c-axis is approximately 7 times larger than that along the a- or b-axis, which contributes to the material's remarkable anisotropic thermoelectric properties [7–10]. The maximum electrical conductivity is approximately 10^3 S/cm. The electrical conductivity along the a,b-plane is approximately three times larger than that along the c-axis [6]. The lattice thermal conductivity along the a,b-plane (1.5 W/(m·K)) is approximately two times larger than that along the c-axis (0.7 W/(m·K)) [11]. The Seebeck coefficient of Bi_2Te_3 almost does not exhibit anisotropy [7]. The Seebeck coefficient varies from -200 to 200 μV/K upon addition of acceptor or donor impurities. In order to reduce the lattice thermal

M. Takashiri (✉)
Department of Materials Science, Tokai University, Hiratsuka, Kanagawa, Japan
e-mail: takashiri@tokai-u.jp

© Springer Nature Switzerland AG 2019
P. Mele et al. (eds.), *Thermoelectric Thin Films*,
https://doi.org/10.1007/978-3-030-20043-5_1

Fig. 1.1 Schematic crystal
structure of Bi_2Te_3

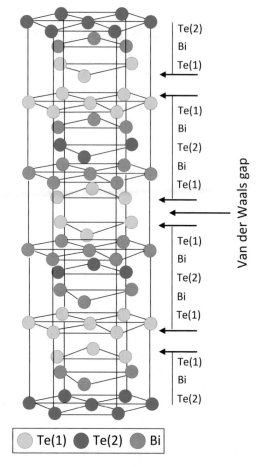

conductivity, ternary alloys such as $(Bi_{1-x}Sb_x)_2Te_3$ and $Bi_2(Se_{1-x}Te_x)_3$ are prepared
[12, 13]. $(Bi_{1-x}Sb_x)_2Te_3$ is formed by replacing Sb atoms with Bi atoms, and exhibits
a p-type conduction behavior. On the other hand, $Bi_2(Se_{1-x}Te_x)_3$ is formed by
replacing Se atoms mainly with $Te^{(2)}$ atoms, and exhibits an n-type conduction
behavior.

Figure 1.2 shows the band structure of Bi_2Te_3 determined by a first-principle
calculation [14]. It should be noted that Bi_2Se_3, Sb_2Te_3, and ternary alloys exhibit
similar band structures [15–17]. Even though the band structure slightly changes

Table 1.1 Lattice constants
and bandgaps of
Bi_2Te_3-based materials [6]

	Lattice constant		Band gap
	a,b-axis	c-axis	
	Å	Å	eV
Bi_2Te_3	4.35	30.3	0.13
Sb_2Te_3	4.25	30.3	0.22
Bi_2Se_3	4.14	28.6	0.12

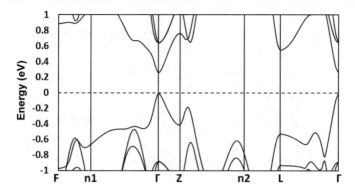

Fig. 1.2 Band structure of Bi_2Te_3

with the calculation parameters, Bi_2Te_3 exhibits properties of an indirect-transition-type semiconductor with a band gap of 0.1–0.2 eV. Its complex crystal structure yields a valence band maximum in the middle of the Z–F direction and conduction band minimum in the middle of the Γ–Z or Z–F direction. Upon carrier doping, Bi_2Te_3 exhibits a multi-valley structure.

1.2 Thin-Film Deposition Methods

1.2.1 Conventional Deposition Methods for Bi₂Te₃-Based Alloy Thin Films

Various film deposition methods including sputtering [18–20], vacuum evaporation [21–24], electrodeposition [25–28], drop-casting [29–32], pulsed laser deposition [33–36], metal organic chemical vapor deposition (MOCVD) [37–39], and molecular beam epitaxy (MBE) [40–42] have been used to deposit Bi_2Te_3-based alloy thin films. As mentioned above, the advantages of the thin-film technology are the reductions in device size and manufacturing cost by incorporating nanoscale effects. Among the various film deposition methods, sputtering, electrodeposition, and printing are the most favorable to realize the advantages of the film deposition. Therefore, we introduce these three deposition methods in this section.

1.2.2 Sputtering Deposition Method

In sputtering, high-energy particles are incident on a target material; they remove constituent atoms of the target from its surface, which are then deposited as a thin film on a substrate. Generally, the sputtering deposition methods are divided

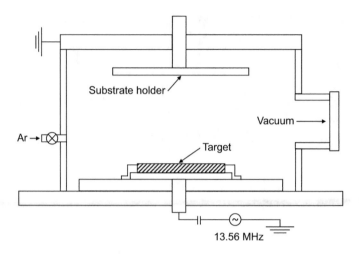

Fig. 1.3 Schematic diagram of the RF magnetron sputtering system

into direct-current (DC) sputtering and radio-frequency (RF) sputtering. The comportments in the vacuum chamber of both DC and RF sputtering are basically the same. The difference is the application of a DC or RF power between the substrate and target to generate the plasma. One of the advantages of the RF sputtering is the ability to use insulator targets and deposit insulating films and conductive oxide films. In addition, to increase the deposition rate and decrease the thin film damage due to high-energy anion impacts, a magnetic field is introduced into the RF sputtering. Figure 1.3 shows a schematic diagram of the RF magnetron sputtering equipment. By installing a permanent magnet on the back of the target, a parallel leakage magnetic field from the target surface to the center is generated. Owing to the parallel magnetic field on the front face of the target, the secondary electrons emitted from the target surface drift on the target owing to the Lorentz force, enabling to efficiently organize the ionization effect even at a low discharge gas pressure. Therefore, an improvement in deposition rate by a high-current-density discharge could be achieved. In addition, to further improve the thermoelectric performance, one approach is to increase the electrical conductivity through an increase in the number of charge carriers in the films by adding H_2 gas during the sputtering (in general, pure Ar gas atmospheres are used) [43–46]. It has been reported that the electrical conductivities of transparent conducting oxide (TCO) films, including indium tin oxide (ITO) and doped zinc oxide (ZnO), increased with the introduction of H_2 gas.

We present typical Bi_2Te_3 thin films prepared using RF magnetron sputtering with the introduction of H_2 gas [47]. The surface morphologies analyzed by scanning electron microscope (SEM) of Bi_2Te_3 thin films with different mixing ratios are shown in Fig. 1.4. The surface morphologies of the thin films were strongly dependent on the mixing ratio. The thin film prepared with pure Ar had a relatively rough surface with densely arranged irregular crystal grains with sharp

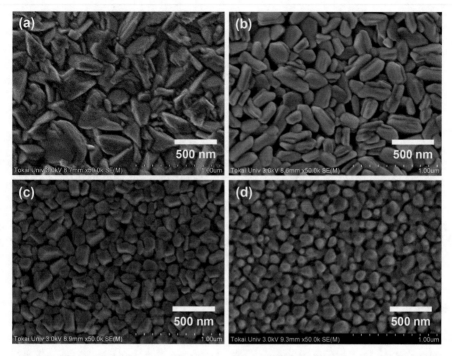

Fig. 1.4 SEM images showing the surface morphologies of the Bi_2Te_3 thin films. (**a**–**d**) Surface morphologies of the samples with $H_2/(H_2 + Ar)$ mixing ratios of 0%, 5%, 10%, and 15%, respectively [47]

edges (Fig. 1.4a). At a mixing ratio of 5% (Fig. 1.4b), hexagonal plate-like crystal grains with a plate thickness of ~300 nm and diameter of ~100 nm were randomly arranged, and voids were observed between the grains. When the mixing ratio was further increased to 10% (Fig. 1.4c), isotropic granular grains with sizes of ~100 nm were densely arranged. The increase in the mixing ratio to 15% (Fig. 1.4d) yielded slightly smaller rounded crystal grains, compared to those observed for the mixing ratio of 10%; the sizes of the voids between the grains slightly increased. These changes in surface morphology implied that surface atoms, particularly those located at the edges of the grains, evaporated when the proportion of H_2 gas increased.

Figure 1.5 shows the atomic compositions of the thin films with different $H_2/(H_2 + Ar)$ ratios, determined by electron probe microanalyzer (EPMA). The atomic concentrations of tellurium and bismuth varied with the mixing ratio. The samples with mixing ratios of 0%, 5%, and 10% exhibited approximately stoichiometric proportions. However, when the mixing ratio increased to 15%, the concentration of bismuth increased, while that of tellurium decreased. The atomic composition was: Bi: 56 at.% and Te: 44 at.%, which significantly deviated from the stoichiometric proportion (Bi: 40 at.% and Te: 60 at.%). Considering both deposition rate and atomic composition, Te atoms were mainly evaporated from the

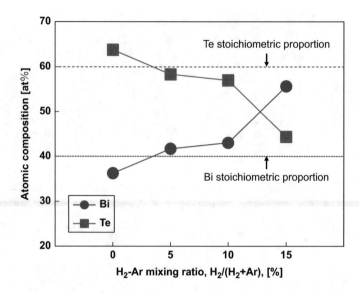

Fig. 1.5 Atomic concentrations of bismuth and tellurium as a function of the $H_2/(H_2 + Ar)$ mixing ratio [47]

film surface by a chemical reaction with hydrogen, producing hydrogen telluride [48, 49].

Figure 1.6 shows relationships between $H_2/(H_2 + Ar)$ and in-plane thermoelectric properties of the Bi_2Te_3 thin films. The Seebeck coefficient was significantly influenced by the proportion of H_2 gas, as shown in Fig. 1.6a. The Seebeck coefficient of the Bi_2Te_3 thin film at a mixing ratio of 0% was -126 $\mu V/K$; when the mixing ratio was increased, the absolute value decreased. In particular, the Seebeck coefficient significantly decreased at a mixing ratio of 15%, as the atomic concentrations of Bi and Te considerably deviated from the stoichiometric proportions of Bi_2Te_3, and thus, the BiTe phase appeared [50]. Figure 1.6b shows that the electrical conductivities of the Bi_2Te_3 thin films prepared with the mixing ratios of 0% and 5% were 351 and 429 S/cm, respectively. The latter value was not the highest value of all of the samples even though this thin film had the largest crystallite size and highest degree of crystal orientation, possibly owing to the appearance of voids between the grains and oxygen concentration inside the film higher than those inside the films prepared at the mixing ratios of 10% and 15%. For the mixing ratio of 10%, the electrical conductivity increased to 741 S/cm. This increase was thought to be determined by two factors. The first factor was a reduction in the oxygen concentration inside the thin film contributing to an increase in mobility or suppression of the decrease in mobility due to the lower crystal orientation [51]. The other factor was the appearance of Te vacancies acting as donors, leading to an increase in the carrier concentration [52, 53]. We considered that the appearance of the Te vacancies contributes more than the reduction in the

Fig. 1.6 (**a**) Seebeck coefficients, (**b**) electrical conductivities, and (**c**) power factors of the Bi_2Te_3 thin films as a function of the $H_2/(H_2 + Ar)$ mixing ratio [47]

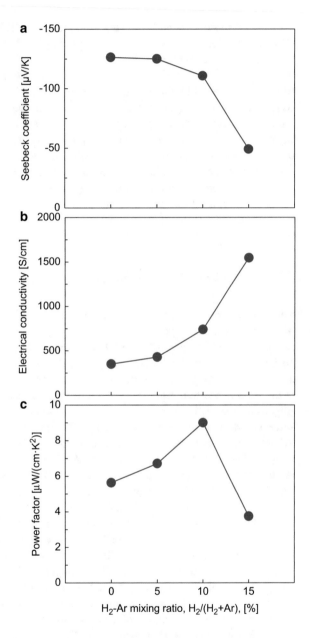

oxygen concentration, as the variation in the content of tellurium is larger than that of oxygen. An excessive thermal annealing has caused a similar phenomenon [54, 55]. When the mixing ratio was further increased, the electrical conductivity significantly increased. The thin film prepared at a mixing ratio of 15% exhibited the highest electrical conductivity of 1547 S/cm, even though it exhibited the smallest crystallite size and voids between the grains. This phenomenon can also be explained by the appearance of the BiTe phase [50]. The power factor as a function of the H_2:Ar mixing ratio is shown in Fig. 1.6c. The power factor of the thin film at the mixing ratio of 0% was relatively low, 5.6 $\mu W/(cm \cdot K^2)$, which was attributed to the lower electrical conductivity owing to the smaller crystallite size and higher oxygen concentration, despite the relatively high Seebeck coefficient. The power factor increased with the mixing ratio. At the mixing ratio of 10%, the power factor reached the highest value of 9.0 $\mu W/(cm \cdot K^2)$, owing to the relatively high electrical conductivity and Seebeck coefficient. When the mixing ratio was 15%, the power factor was significantly decreased owing to the decrease in the Seebeck coefficient. Therefore, a moderate mixing ratio (10%) of H_2 gas promoted a low oxygen concentration in the thin film without causing voids, leading to the higher electrical conductivity and power factor.

1.2.3 Electrodeposition Methods

The electrodeposition method is one of the wet processes in a solution phase. There are mainly two types of electrodepositions: electrolytic and non-electrolytic platings. The electrodeposition method generally has advantages such as a lower cost, compared with the dry process techniques. In addition, the electrodeposition is a low-temperature process below 80 °C. Another advantage of the electrodeposition is the high deposition rate, which enables to deposit a thick film in a short time. Regarding the Bi_2Te_3-based alloy thin films, the electrolytic plating has been used since Takahashi et al. electrodeposited Bi_2Te_3-based alloy films for the first time in the 1990s [56]. Figure 1.7 presents a schematic diagram of the electrolytic plating method. The thin films are generally prepared using a standard three-electrode cell system, with working, counter, and reference electrodes inserted in an electrolyte solution. The thin film is deposited on the working electrode when a constant voltage (potentiostatic electrodeposition) or constant current density (galvanostatic electrodeposition) is applied between the working and counter electrodes.

We present typical Sb_2Te_3 thin films prepared using potentiostatic electrodepositions [57]. In order to increase the film crystallinities, thermal annealing at different temperatures was performed. The SEM images in Fig. 1.8 show that the surface morphologies of the electrodeposited Sb_2Te_3 films are strongly affected by the annealing temperature. Figure 1.8a shows that the as-deposited film is covered with approximately spherical grains of a smooth surface; the grain size is approximately 1 μm. When the thin film is annealed at 200 °C, the shape of the grains remains unchanged, but the grain size increases to approximately 2 μm (Fig. 1.8b). The grain shape and size of the thin film annealed at 250 °C are almost the same as those of the thin film annealed at 200 °C (Fig. 1.8c). At an annealing temperature of 300 °C, the grain size remains unchanged from that of the thin film at 250 °C, but the grain surface becomes confetti-like (Fig. 1.8d). The surface of the thin film annealed at 350 °C seems to have needle-shaped crystals growing on the spherical

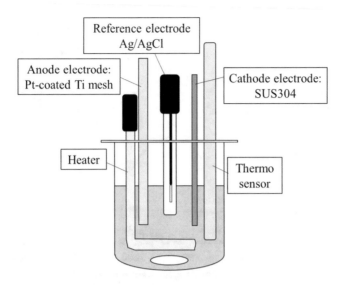

Fig. 1.7 Schematic diagram of the electrodeposition system

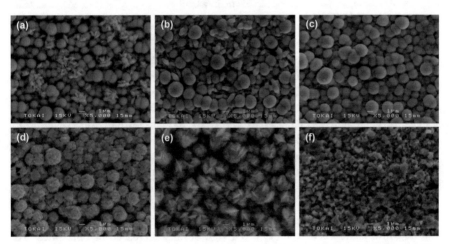

Fig. 1.8 SEM images of the surface morphologies and grain structures of the Sb_2Te_3 thin films annealed at (**a**) room temperature (as-deposited), (**b**) 200, (**c**) 250, (**d**) 300, (**e**) 350, and (**f**) 400 °C [57]

grains, with an increased space between grains (Fig. 1.8e). Furthermore, when the annealing temperature increases to 400 °C, the surface morphology significantly changes (Fig. 1.8f). The thin film exhibits a porous structure with irregularly shaped submicron grains.

The dependence of the atomic composition on the annealing temperature and impurities originated from the SUS304 stainless-steel substrate (Fe:Cr:Ni = 70:19:9) is shown in Fig. 1.9. It should be noted that the EMPA

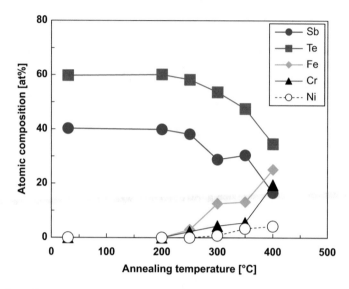

Fig. 1.9 Annealing temperature dependences of the atomic compositions of the Sb$_2$Te$_3$ thin films, determined by EPMA [57]

measurement was performed after the thin films were removed from the substrates. The as-deposited thin-film data are plotted at 27 °C. The as-deposited film and the film annealed at 200 °C exhibit the stoichiometric atomic composition (Sb:Te = 40:60); no contamination from the substrate is observed. When the annealing temperature is increased to 250 °C, the atomic concentrations of tellurium and antimony decreased, and small amounts of iron and chromium can be observed. This indicates that the elements in the substrate gradually diffuse into the Sb$_2$Te$_3$ thin film upon the increase in the annealing temperature. With the further increase in the annealing temperature to 300 °C, the atomic concentrations of tellurium and antimony continue to decrease, and Fe, Cr, and Ni impurities are detected. In particular, the atomic concentration of Fe reaches approximately 13 at.%. Finally, the thin film annealed at 400 °C has a significantly increased of impurities. The total atomic concentration of impurities from the substrate reaches approximately 50 at.%.

Figure 1.10 shows relationships between the annealing temperature and in-plane thermoelectric properties of the Sb$_2$Te$_3$ thin films. In Fig. 1.10a, the as-deposited thin film and thin film annealed at 200 °C with relatively high Seebeck coefficients of approximately 170 μV/K. In general, antimony telluride alloys exhibit high Seebeck coefficients at their stoichiometric atomic compositions. With the increase in the annealing temperature, the Seebeck coefficient decreases. The Seebeck coefficient of the thin film annealed at 350 °C is 66 μV/K. At the annealing temperature of 400 °C, the Seebeck coefficient −37 μV/K, indicating that the thin film has become an n-type semiconducting material, as the Sb$_2$Te$_3$ crystal structure does not exist in the thin film at this annealing temperature. In Fig. 1.10b, the

Fig. 1.10 Annealing temperature dependences of the (**a**) Seebeck coefficients, (**b**) electrical conductivities, and (**c**) power factors of the Sb$_2$Te$_3$ thin films [57]

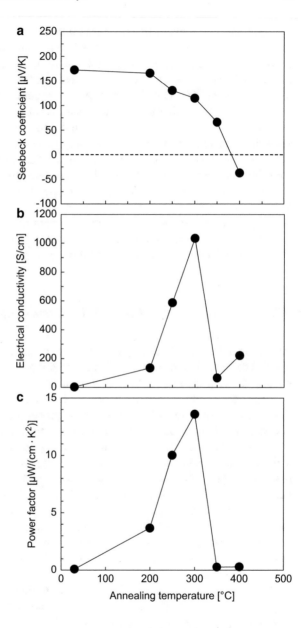

as-deposited thin film exhibits a very low electrical conductivity, possibly as this film has a lower mobility owing to the low crystallinity and small crystallite size, whereas the carrier concentration is suitable leading to the high Seebeck coefficient. The electrical conductivity rapidly increases between the annealing temperatures of 200 and 300 °C. The thin film annealed at 300 °C exhibits the maximum electrical conductivity of 1034 S/cm. The rapid increase in the electrical conductivity is due to the crystallinity improvement and increase in the carrier concentration induced by the impurity contamination. At the annealing temperature of 350 °C, the electrical conductivity rapidly decreases, possibly as the gaps between the grains increase, as shown in Fig. 1.9e. The lowest electrical conductivity is observed for the thin film annealed at 400 °C owing to its porous structure, as shown in Fig. 1.9f. We calculated the power factor using the measured Seebeck coefficient and electrical conductivity, as shown in Fig. 1.10c. The dependence of the power factor on the annealing temperature has a similar trend as that of the electrical conductivity. We obtain antimony telluride thin films with higher thermoelectric performances at the annealing temperatures of 250 and 300 °C. The thin film annealed at 300 °C exhibits the maximum power factor of 13.6 $\mu W/(cm \cdot K^2)$. As mentioned above, the thin film annealed at 300 °C contains a certain amount of impurities from the substrate. The impurities contribute to the decrease in the Seebeck coefficient as well as to the increase in the electrical conductivity. The contribution to the increase in the electrical conductivity surpasses that to the decrease in the Seebeck coefficient.

1.2.4 Combination Method of Sputtering and Electrodeposition

As mentioned above, electrodeposition is one of the most favorable methods as it is very cost-effective owing to its simple scalability and low operating temperature without the requirement for vacuum conditions. However, one of the drawbacks of electrodeposition is that films are deposited only on conductive electrodes. Therefore, to measure their thermoelectric properties or fabricate thermoelectric devices, the films should be transferred from conductive to insulating substrates using resin adhesives. During this transfer, micropores appear in the films, hindering their electrical conductions. Therefore, a method that can suppress the micropores in the films should be developed. A possible method is to electrochemically deposit the films on seed layers, thus generating films with the same compositions as those of the electrodeposited films using a dry process such as sputtering on insulating substrates [58]. The thermoelectric properties of these electrodeposited films can be measured without the transfer process. Furthermore, this method is expected to promote the crystal growth of electrodeposited films owing to the formation of a homogeneous interface.

We present typical as-grown Bi_2Te_3 electrodeposited films using sputtered Bi_2Te_3 seed layers (AES films). Figure 1.11 shows the SEM surface and cross-section morphologies of a typical AES film. Figure 1.11a shows that the AES film is composed of nanoscale (smaller than 0.5 μm) crystal grains with a densified morphology and smooth and homogeneous surface. A cross-section image of the AES film is shown in Fig. 1.11b. The interface between the electrodeposited film and seed layer is clearly observed; no voids are observed. The grain size of the electrodeposited film was larger than that of the seed layer. The measured thicknesses of the electrodeposited film and seed layer were 1.6 and 1.3 μm, respectively. For comparison, we analyzed the surface morphologies of the standard electrodeposited

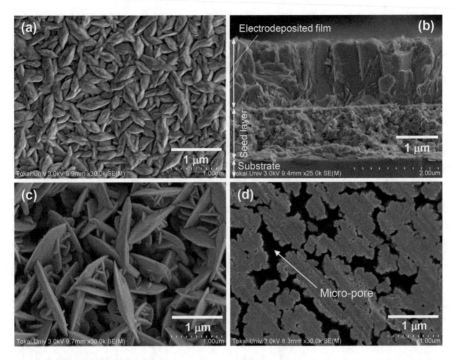

Fig. 1.11 SEM images showing the morphologies of the Bi_2Te_3 films. (**a, b**) Surface and cross-section images of the AES film, respectively; (**c, d**) top and bottom surface images of the standard electrodeposited film, respectively [58]

film. The top surface image (Fig. 1.11c) shows dendritic growths on the film surface, leading to the rough appearance of the surface. Figure 1.11d shows the bottom surface of the film after detachment from the stainless-steel substrate. A large area of micropores was observed between the crystal grains. Consequently, we attributed the lower electrical conductivity and lower absolute Seebeck coefficient of the standard electrodeposited film to the presence of micropores.

Figure 1.12 shows the in-plane thermoelectric properties of the Bi_2Te_3 films as a function of the applied current density. As a reference, the thermoelectric properties of Bi_2Te_3 films obtained by the standard electrodeposition are also presented in the figure. The reference samples were formed on a stainless-steel substrate using potentiostatic electrodeposition, and subsequently detached from the stainless-steel substrate using epoxy resin. As shown in Fig. 1.12a, the electrical conductivity of the AES film at an applied current density of 1.8 mA/cm^2 was 646 S/cm. When the current density was increased to 2.1 mA/cm^2, the electrical conductivity of the AES film increased, reaching the highest value of 674 S/cm. With the further increase in the current density, the electrical conductivity decreased. The electrical conductivity of the AES film deposited at an applied current density of 2.6 mA/cm^2 was 523 S/cm. The electrical conductivities of the electrodeposited films followed a similar trend with that of the AES films; however, the electrical

Fig. 1.12 (**a**) Electrical conductivities, (**b**) Seebeck coefficients, and (**c**) power factors of the Bi_2Te_3 films, including the AES film, electrodeposited film, and reference standard electrodeposited film, as a function of the applied current density [58]

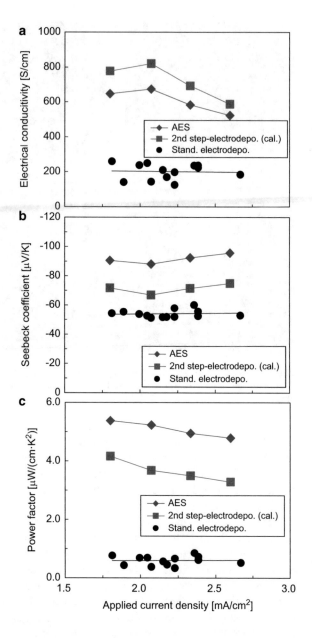

conductivity magnitudes of the electrodeposited films were higher than those of the AES films. This indicates that the electrical conductivity of the electrodeposited film was higher than that of the seed layer. The highest value of 821 S/cm was observed for the electrodeposited film obtained at a current density of 2.1 mA/cm^2. Furthermore, the electrical conductivity of the standard electrodeposited film was approximately 200 S/cm, in the range of the applied current density. Therefore, the electrodeposition using a seed layer contributed to the improvement in the electrical conductivity. Figure 1.12b shows that the Seebeck coefficient of the AES film at an applied current density of 1.8 mA/cm^2 is -91 μV/K. When the current density was increased, the Seebeck coefficient of the AES film did not significantly change. The Seebeck coefficient of the AES film at an applied current density of 2.6 mA/cm^2 was -96 μV/K. To the best of our knowledge, this Seebeck coefficient is significantly higher than the previously reported values. The Seebeck coefficients of the electrodeposited films exhibited a similar trend with that of the AES films. The Seebeck coefficients of the electrodeposited films were in the range of -67 to -75 μV/K. The absolute values were approximately 25% lower than those of the AES films. The standard electrodeposited film exhibited a Seebeck coefficient of approximately -50 μV/K. Therefore, the electrodeposition using the seed layer also improved the Seebeck coefficient. The power factor as a function of the applied current density is shown in Fig. 1.12c. The standard electrodeposited film exhibited a power factor of approximately 0.5 μW/(cm·K^2) within the applied current density range. Moreover, the power factor of the AES film at an applied current density of 1.8 mA/cm^2 was 5.4 μW/(cm·K^2). With the increase in the current density, the power factor of the AES film linearly decreased. For the electrodeposited film, the power factor at an applied current density of 1.8 mA/cm^2 was 4.2 μW/(cm·K^2). With the increase in the current density, the power factor linearly decreased with approximately the same slope as that of the AES films. Overall, the power factors of the electrodeposited films were approximately 15% lower than those of the AES films, indicating that the power factors of the electrodeposited films were lower than that of the seed layer. However, the power factors of the electrodeposited films were 7–9 times higher than that of the standard electrodeposited film. The maximum power factor in this study is comparable to the excellent values of as-grown electrodeposited Bi–Te–Se ternary compounds [59, 60]. In summary, the electrodeposition using the seed layer significantly improved the power factor mainly owing to the increase in the electrical conductivity.

1.2.5   Printing Method

Unlike other film deposition methods such as the dry and wet processes, the printing method uses nanoparticle inks, which are applied to a substrate. This method is the most convenient procedure for the preparation of thin layers with fine patterns. There are mainly two types of printing methods: screen printing and inkjet printing [61–63]. Screen printing is a kind of stencil printing. It is not to print with a plate coated with ink, but to print squeezing ink out of holes made on a plate itself. Inkjet printing is a type of printing that propels drops of ink onto a substrate. One of the important factors in the printing methods is the quality of the ink including thermoelectric nanoparticles. The size and shape of the nanoparticles affect the thermoelectric performance [64–66]. For the analysis of the thermoelectric properties of printed thin films, the drop-cast method is the most appropriate as the thin film can be formed simply by dropping the ink on the substrate and drying it.

The methods for production of nanoparticles can be roughly divided into two types. One of them is a top-down method, in which bulk solids are physically transformed into nanoparticles by pulverization [29, 67]. The other type is a bottom-up method of synthesis of atoms and molecular aggregates (crystals, amorphous particles) through atomic- and molecular-level chemical reactions [68–70].

A common bottom-up synthesis method is ball-milling or bead-milling, based on the most basic crushers used for fine pulverizations of powders and powder raw material particles. Figure 1.13 shows a schematic diagram of the ball-milling method. In this method, the powder material, which should be pulverized, and hard balls, such as metal (stainless steel or tungsten) balls, are placed in a crushing vessel (pot). The pot is placed on the roller of the ball mill equipment and rotates them in the direction of rotation. The raw material powders are pulverized more finely by gradually grinding them by collisions and frictions between the balls in the pot. SEM micrographs of $Bi_{0.4}Te_{3.0}Sb_{1.6}$ nanoparticles fabricated using bead-milling are shown in Fig. 1.14 [32]. The average size of the nanoparticles is approximately 50 nm. The selected area electron diffraction (SAED) pattern of the nanoparticles in Fig. 1.15 shows a dotted structure of diffraction rings with bright spots [29]. This indicates that the nanoparticles exhibit a polycrystalline structure.

We present typical nanoparticle ($Bi_{0.4}Te_{3.0}Sb_{1.6}$) thin films prepared using the drop-casting method [32]. In order to connect the nanoparticles to each other,

Fig. 1.13 Schematic diagram of the ball-milling method

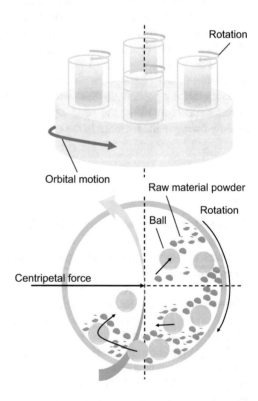

Fig. 1.14 SEM micrograph of the $Bi_{0.4}Te_{3.0}Sb_{1.6}$ nanoparticles [32]

Fig. 1.15 Electron diffraction pattern of the $Bi_{0.4}Te_{3.0}Sb_{1.6}$ nanoparticles [32]

thermal annealing at different temperatures was performed. The cross-section and surface micrographs of the thin films shown in Fig. 1.16 reveal the effects of the annealing temperature. The thin film annealed at 300 °C contains only nanoparticles, which essentially have the same sizes as those in Fig. 1.14. At the annealing temperature of 350 °C, crystal flakes with hexagonal shapes with widths and thicknesses of approximately 500 nm and 50 nm, respectively, were formed on the surface of the thin film. With the further increase in the annealing temperature, the crystal flakes on the surface became larger. The crystal flakes on the nanoparticles exhibited single-crystalline structures [29]. The in-plane thermoelectric properties of the nanoparticle thin films measured at approximately 300 K are shown in Fig. 1.17. At the annealing temperature of 300 °C, the Seebeck coefficient and electrical conductivity of the thin film are 258 μV/K and 1.3 S/cm, respectively. The electrical conductivity is low as the organic components such as

Fig. 1.16 Cross-section and surface SEM micrographs of the $Bi_{0.4}Te_{3.0}Sb_{1.6}$ nanoparticle thin films sintered at various temperatures [32]

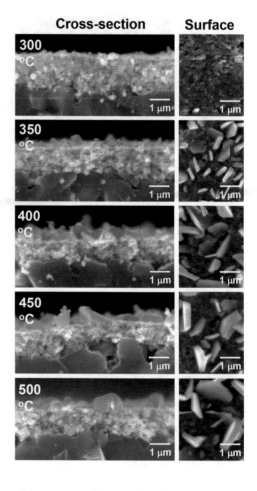

the surfactant remain on the surfaces of the nanoparticles so that electrons cannot pass through the surfactant layer between the nanoparticles. With the increase in the annealing temperature, the Seebeck coefficient of the thin film decreased, while the electrical conductivity increased. At an annealing temperature of 500 °C, the Seebeck coefficient and electrical conductivity of the thin film are 65 μV/K and 109 S/cm, respectively. The highest achieved thermoelectric power factor is 1.3 μW/(cm·K^2) at an annealing temperature of 350 °C. Compared to that of the bulk material, the thermoelectric power factors of the nanoparticle thin films are low. In order to further improve the thermoelectric properties of the nanoparticle thin films, it is necessary to investigate in detail the origin of the degraded thermoelectric properties.

The bottom-up synthesis method generally utilizes a homogeneous nucleation followed by a solid-phase precipitation. For the synthesis of thermoelectric nanoparticles including nanoplates, nanosheets, nanorods, etc., liquid-phase methods (hydrothermal method, liquid-phase precipitation method, and electro-chemical synthesis), gas-phase methods (chemical vapor deposition and chemical

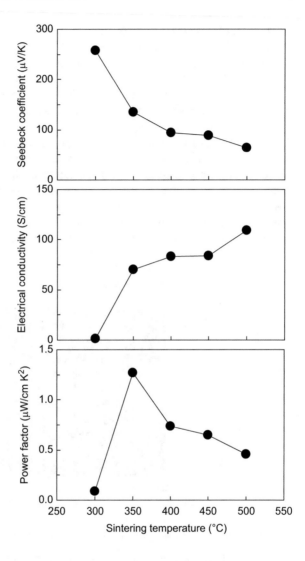

Fig. 1.17 In-plane Seebeck coefficients, electrical conductivities, and thermoelectric power factors of the $Bi_{0.4}Te_{3.0}Sb_{1.6}$ nanoparticle thin films as a function of the annealing temperature [32]

vapor transport), and solid-phase method (spinodal decomposition method) are commonly employed [71–73]. Among them, the hydrothermal method can be used to effectively fabricate nanoscale single-crystals, which are very favorable for a thermoelectric performance improvement. Therefore, the hydrothermal synthesis will be described below as a typical example of a bottom-up synthesis method.

The hydrothermal synthesis is a synthesis or crystal growth of a compound performed in the presence of hot water at high temperature and pressure. As substances insoluble in water are easily dissolved at normal temperature and pressure, it is possible to synthesize and grow substances that normally cannot be obtained. In particular, a hydrothermal synthesis method using an organic solvent instead of hot water is referred to as solvothermal method. In the bismuth tellurium system, it is common to use the solvothermal method. Figure 1.18 shows a schematic diagram of the hydrothermal synthesis. The starting material and water (organic solvent) are usually placed in a sealed container (autoclave); the container is sealed and heated to obtain the product. After the synthesis, the products are naturally cooled down to room temperature. The products were collected using centrifugation and were washed several times in distilled water and absolute methanol. Finally, they were dried in vacuum at a temperature lower than 100 °C for approximately 24 h. A transmission electron microscope (TEM) image of a typical Bi_2Te_3 nanoplate synthesized using the solvothermal method is shown in Fig. 1.19 [66]. The shape of the nanoplate is a regular hexagon, whose surface is significantly flattened. The edge size of the nanoplate was approximately 900 nm and the thickness was expected to be very small (smaller than 50 nm) as the mesh structure behind the nanoplate can be observed. The SAED pattern shown in the inset of Fig. 1.19 was obtained from the tip of the hexagonal nanoplate. The hexagonally

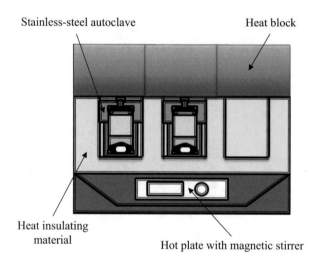

Stainless-steel autoclave   Heat block

Heat insulating material   Hot plate with magnetic stirrer

Fig. 1.18 Schematic diagram of the hydrothermal synthesis [31]

Fig. 1.19 TEM image of a Bi_2Te_3 nanoplate prepared by a solvothermal synthesis. The inset shows the corresponding SAED pattern [66]

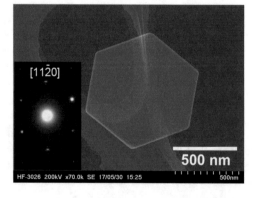

symmetric spot pattern indicated single crystallinity and can be indexed on the basis of a rhombohedral cell. Briefly, the mechanism of crystal growth of Bi_2Te_3 nanoplates is as follows. Once the Bi_2Te_3 nuclei were generated by the chemical reactions, they aggregated. When the radius of the aggregated nuclei became larger than that of their critical nucleus, Bi_2Te_3 nanoparticles were generated. The formation of the nanoplate was also attributed to the inherent crystal structure. Owing to the high surface energy of the nuclei, the aggregated Bi_2Te_3 particles were not in thermodynamic equilibrium and were metastable; therefore, Bi_2Te_3 nanoplates with better crystallinities became thermodynamically preferred. After the formation of Bi_2Te_3 nanoplates, the Ostwald ripening process proceeded. The smaller nanoparticles adsorbed around plates were consumed and grew gradually leading to larger nanoplates along the top–bottom crystalline plane. This occurred as the rhombohedral Bi_2Te_3 can be described as a stack of infinite layers extending along the top–bottom crystalline plane connected by van der Waals bonds, as shown in Fig. 1.1. From a thermodynamic perspective, the free energy of a broken covalent bond is higher than that of a dangling van der Waals bond. This implies that the Bi_2Te_3 crystal growth along the top–bottom crystalline plane should be significantly faster than that along the c-axis, which is perpendicular to the top–bottom planes, as the crystalline facets tend to develop on the low-index plane.

Typical nanoplate (Bi_2Te_3) thin films were prepared using the drop-casting method [66]. In order to connect the nanoplates to each other, thermal annealing at different temperatures was performed. Cross-section and surface micrographs of the as-prepared thin film are shown in Fig. 1.20. The surface image in Fig. 1.20a shows that the thin film is composed of high-purity hexagonal nanoplates with edge sizes in the range of 500–2000 nm (average size of 1000 nm). The hexagonal nanoplates were well aligned in the direction of the film surface, but some apertures were observed in the thin film. A cross-sectional image of the nanoplate thin film is shown in Fig. 1.20b. The hexagonal Bi_2Te_3 nanoplates were well piled up; the approximate thickness of the thin film was 40 µm. Figure 1.21 shows SEM images of the surface morphologies of the Bi_2Te_3 nanoplate thin films obtained at different

Fig. 1.20 (a) Surface and (b) cross-sectional morphologies of an untreated Bi_2Te_3 nanoplate thin film, observed by SEM [66]

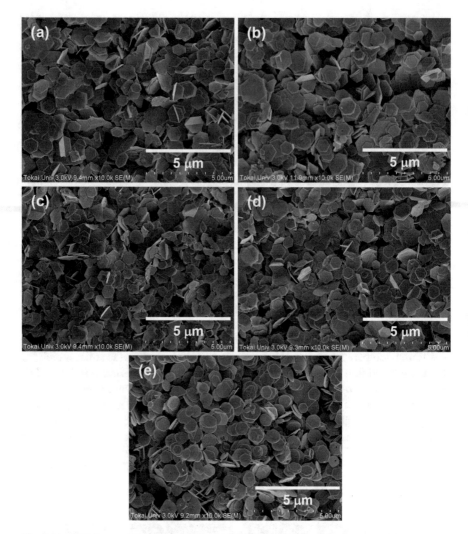

Fig. 1.21 SEM images of the surface morphologies of the Bi_2Te_3 nanoplate thin films annealed at (**a**) 200, (**b**) 250, (**c**) 300, (**d**) 350, and (**e**) 400 °C [66]

annealing temperatures. Overall, in all of the Bi_2Te_3 nanoplate thin films obtained at annealing temperatures of 200–400 °C, high-purity nanoplates were well aligned in the direction of the film surface. In terms of fine structure, the nanoplate thin films annealed at 200 and 250 °C clearly exhibited the edges of the hexagonal nanoplates (Figs. 1.21a, b). However, with the increase in the annealing temperature to 300 and 350 °C, the edges of the hexagonal nanoplates gradually disappeared (Figs. 1.21c, d). The shape of the nanoplates annealed at 400 °C was not hexagonal, but circular (Fig. 1.21e), as Bi or Te atoms at the edges of the hexagonal nanoplates evaporated during the thermal annealing at the higher temperature.

The in-plane thermoelectric properties of the nanoplate thin films measured at approximately 300 K are shown in Fig. 1.22. The thermoelectric properties of two samples, untreated thin film and thin film annealed at 200 °C, are not included in Fig. 1.22 as their electrical resistivities were too high so that the thermoelectric properties could not be measured. This indicates that the nanoplates did not tightly join with each other when the annealing temperature was smaller than 200 °C.

Fig. 1.22 In-plane thermoelectric properties of the Bi_2Te_3 nanoplate thin films as a function of the annealing temperature. (**a**) Seebeck coefficient, (**b**) electrical conductivity, and (**c**) power factor [66]

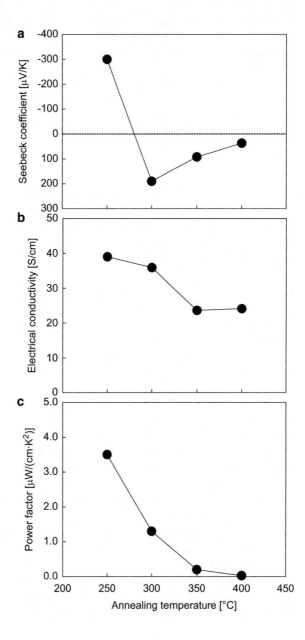

Figure 1.22a shows that the Seebeck coefficient of the nanoplate thin film annealed at 250 °C was $-300 \, \mu V/K$, higher than those reported for Bi_2Te_3 materials [6]. With the increase in the annealing temperature, the Seebeck coefficient became positive. The nanoplate thin film annealed at 300 °C exhibited a Seebeck coefficient of $190 \, \mu V/K$. With the increase in the annealing temperature, the Seebeck coefficients of the nanoplate thin films approached the value of zero. This occurred as the atomic composition of the thin film was different from the stoichiometric composition owing to the partial atomic evaporation. Figure 1.22b shows that the electrical conductivity of the nanoplate thin film annealed at 250 °C was 39 S/cm, which is the highest value in this study. This indicates that the contact between nanoplates was enhanced by the thermal annealing. The annealing temperature dependence of the power factor of the nanoplate thin films is shown in Fig. 1.22c. The power factor of the nanoplate thin film rapidly decreased with the increase in the annealing temperature. The maximum power factor of $3.5 \, \mu W/(cm \cdot K^2)$ was obtained at the annealing temperature of 250 °C.

1.3   Thin-Film Thermoelectric Generators

Typical thin-film thermoelectric generators prepared by RF magnetron sputtering are presented in this section. The fabrication process of the multi-layered thermoelectric generator is illustrated in Fig. 1.23 [3]. Initially, we prepared a glass substrate (Eagle XG, Corning) ($20 \times 30 \, mm^2$; thickness: 0.3 mm), thinner than that used in the experiments to understand the optimal conditions for thermal annealing, for fabrication of as thin as possible generators. A p-type Sb_2Te_3 film with a thickness of 1.0 mm was deposited on one side of the glass substrate, and an n-type film with a thickness of 1.0 mm was sequentially deposited on the other side of the substrate by RF magnetron sputtering. Following the film deposition, a thermal annealing was carried out at the optimal temperature. The p-type film was then connected with the n-type film by spraying silver paste at the one-side edge face of the substrate. After baking the sample to dry the silver paste, the sample was wrapped with an insulation tape, except the part of the opposite edge side of the sample on which silver paste was not sprayed (step 3). Silver paste was then sprayed on the unwrapped positions of both $p-Sb_2Te_3$ and $n-Bi_2Te_3$ films (step 4). In the next step (step 5), a piece of the obtained sample was connected with others as p–n junctions connected in series. In step 6, 11 pieces of samples were connected in series followed by baking of the samples to connect each piece of the sample tightly. Finally, the multi-layered sample was bundled by wrapping the insulation tape. A cross-section photograph of the obtained multi-layered thermoelectric generator shows that the total thickness of the generator is ~7 mm, and that there are no clear gaps between the pieces.

A photograph of the measurement system of the multi-layered thermoelectric generator is shown in Fig. 1.24a. In order to achieve a temperature difference between the ends of the generator, the lower half of the generator was dipped in hot

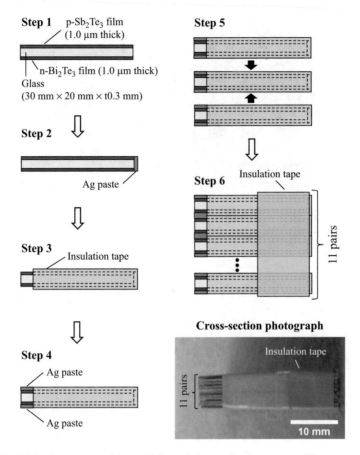

Fig. 1.23 Fabrication process of the multi-layered thermoelectric generator [3]

water, while the upper half was maintained in the atmosphere. The temperatures on both sides were monitored using thermocouples (K-type) attached on the generator. Two electrodes were connected to the Bi_2Te_3 and Sb_2Te_3 sides of the substrate. The temperature difference was generated by changing the temperature of the hot water; the open-circuit voltage (V_{oc}) was measured using a digital multimeter. The performance of the multi-layered thermoelectric generator as a function of the temperature difference is shown in Fig. 1.24b. The temperature difference at the ends was changed from 11 to 28 K. At a temperature difference of 11 K, V_{oc} of 23.7 mV was achieved. With the increase in the temperature difference, V_{oc} linearly increased. V_{oc} of 32.0 mV was achieved at a temperature difference of 28 K. Our results were in relatively good agreement with those in previous studies with similar device structures and temperature differences [74–76]. The maximum output power (P_{max}) is expressed as: $P_{max} = V_{oc}^2/4R_{total}$, where R_{total} is the measured total resistance of the multi-layered thermoelectric generator, which mainly comprises

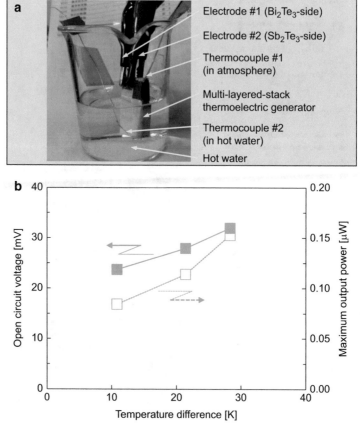

Fig. 1.24 (**a**) Photograph of the measurement system of the multi-layered thermoelectric generator. (**b**) Open-circuit voltages and maximum output powers of the multi-layered thermoelectric generator at various temperature differences [3]

the resistances of the p- and n-type thin films, R_{film}, and contact resistance, R_c. R_{total} of the generator measured using a digital multimeter was 1.6 kΩ. P_{max} was estimated to be 0.08 μW at the temperature difference of 11 K. P_{max} increased with the temperature difference; P_{max} of 0.15 μW was achieved at the temperature difference of 28 K.

1.4 Summary

In this chapter, the crystal structures of Bi_2Te_3-based alloys were presented. An overview of thin-film deposition methods was also presented, and typical conventional deposition methods such as sputtering, electrodeposition, and printing were introduced using typical experimental results of Bi_2Te_3-based alloy thin films. As the presented film fabrication methods have distinct characteristics, the film fabrication method should be selected according to the application. Finally, a typical thin-film thermoelectric generator, consisting of a multi-layered structure of p-type Sb_2Te_3 and n-type Bi_2Te_3 thin films, was introduced. Bi_2Te_3-based alloys are relatively old thermoelectric materials whose research and development began in the 1950s. However, the material properties have been improved using the latest thin-film process technologies by nanostructuring, which makes this research field very attractive with significant potentials for novel findings.

References

1. A. Kadhim, A. Hmood, H. Abu Hassan, Mater. Lett. **97**, 24 (2013)
2. Y. Ito, M. Mizoshiri, M. Mikami, T. Kondo, J. Sakurai, S. Hata, Jpn. J. Appl. Phys. **56**, 06GN06 (2017)
3. K. Takayama, M. Takashiri, Vacuum **144**, 164 (2017)
4. H. Yamamuro, N. Hatsuta, M. Wachi, Y. Takei, M. Takashiri, Coatings **8**, 22 (2018)
5. J.H. Goldsmid, *Materials Used in Semiconductor Devices* (Wiley, New York, 1965)
6. H. Scherrer, S. Scherrer, in *CRC Handbook of Thermoelectrics*, ed. by D. M. Rowe, (CRC Press, Boca Raton, 1995), p. 211
7. H. Kaibe, Y. Tanaka, M. Sakata, I. Nishida, J. Phys. Chem. Solids **50**, 945 (1989)
8. P.J. Taylor, J.R. Maddux, W.A. Jesser, F.D. Rosi, J. Appl. Phys. **85**, 7807 (1999)
9. K. Yamauchi, M. Takashiri, J. Alloys Compd. **698**, 977 (2017)
10. S. Kudo, S. Tanaka, K. Miyazaki, Y. Nishi, M. Takashiri, Mater. Trans. **58**, 513 (2017)
11. G.S. Nolas, J. Sharp, H.J. Goldsmid, *Thermoelectrics* (Springer, New York, 2001)
12. W.M. Yim, F.D. Rosi, Solid State Electron. **15**, 1121 (1972)
13. M. Takashiri, K. Imai, M. Uyama, H. Hagino, S. Tanaka, K. Miyazaki, Y. Nishi, J. Appl. Phys. **115**, 214311 (2014)
14. T. Inamoto, M. Takashiri, J. Appl. Phys. **120**, 125105 (2016)
15. S.K. Mishra, S. Satpathy, O. Jepsen, J. Phys. Condens. Matter **9**, 461 (1997)
16. X. Luo, M.B. Sullivan, S.Y. Quek, Phys. Rev. B Condens. Matter **86**, 184111 (2012)
17. G. Wang, T. Cagin, Phys. Rev. B Condens. Matter **76**, 075201 (2007)
18. K. Yildiz, U. Akgul, H.S. Leipner, Y. Atic, Superlattice. Microst. **58**, 2013 (2013)
19. D. Bourgault, C.G. Garampon, N. Caillault, L. Carbone, J.A. Aymami, Thin Solid Films **516**, 8579 (2008)
20. K. Kusagaya, M. Takashiri, J. Alloys Compd. **653**, 480 (2015)
21. X. Duan, Y. Jiang, Appl. Surf. Sci. **256**, 7365 (2010)
22. A.J. Zhou, L.D. Feng, H.G. Cui, J.Z. Li, G.Y. Jiang, X.B. Zhao, J. Electron. Mater. **42**, 2184 (2013)
23. M. Takashiri, K. Kurita, H. Hagino, S. Tanaka, K. Miyazaki, J. Appl. Phys. **118**, 065301 (2015)
24. M. Uchino, K. Kato, H. Hagino, K. Miyazaki, J. Electron. Mater. **42**, 1814 (2012)
25. J. Kuleshova, E. Koukharenko, X. Li, N. Frety, S. Nandhakumar, J. Tudor, S.P. Beeby, N.M. White, Langmuir **26**, 16980 (2010)
26. K. Matsuoka, M. Okuhata, N. Hatsuta, M. Takashiri, Trans. Mater. Res. Soc. Jpn **40**, 373 (2015)
27. K. Matsuoka, M. Okuhata, M. Takashiri, J. Alloys Compd. **649**, 721 (2015)
28. C.V. Manzano, B. Abad, M.M. Rojo, Y.R. Koh, S.L. Hodson, A.M.L. Martinez, X. Xu, A. Shakouri, T.D. Sands, T. Borca-Tasciuc, M. Martin-Gonzalez, Sci. Rep. **6**, 19129 (2016)
29. M. Takashiri, S. Tanaka, K. Miyazaki, J. Cryst. Growth **372**, 199 (2013)

30. M. Koyano, S. Mizutani, Y. Hayashi, S. Nishino, M. Miyata, T. Tanaka, K. Fukuda, J. Electron. Mater. **46**, 2873 (2017)
31. K. Wada, K. Tomita, M. Takashiri, J. Cryst. Growth **468**, 194 (2017)
32. M. Takashiri, S. Tanaka, M. Takiishi, M. Kihara, K. Miyazaki, H. Tsukamoto, J. Alloys Compd. **462**, 351 (2008)
33. R.S. Makala, K. Jagannadham, B.C. Sales, J. Appl. Phys. **94**, 3907 (2003)
34. H. Obara, S. Higomo, M. Ohta, A. Yamamoto, K. Ueno, T. Iida, Jpn. J. Appl. Phys. **48**, 085506 (2009)
35. P.H. Le, C.N. Liao, C.W. Luo, J. Leu, J. Alloys Compd. **615**, 546 (2014)
36. L. Thi, C. Tuyen, P. Huu, L. Chih, W. Luo, J. Leu, J. Alloys Compd. **673**, 107 (2016)
37. A. Al Bayaz, A. Giani, M.C. Artaud, A. Foucaran, F. Pascal-Delannoy, A. Boyer, J. Cryst. Growth **241**, 463 (2002)
38. H.W. You, S.-H. Bae, J. Kim, J.-S. Kim, C. Park, J. Electron. Mater. **40**, 635 (2011)
39. R. Venkatasubramanian, T. Colpitts, E. Watko, M. Lamvik, N. El-Masry, J. Cryst. Growth **170**, 817 (1997)
40. Y. Kim, A. DiVenere, G.K.L. Wong, J.B. Ketterson, S. Cho, J.R. Meyer, J. Appl. Phys. **91**, 716 (2002)
41. Z. Wang, X. Zhang, Z. Zeng, Z. Zhang, Z. Hua, ECS Electrochem. Lett. **3**, 99 (2014)
42. X. Zhang, Z. Zeng, C. Shen, Z. Zhang, Z. Wang, J. Appl. Phys. **115**, 024307 (2014)
43. K. Zhang, F. Zhu, C.H.A. Huan, A.T.S. Wee, Thin Solid Films **376**, 255 (2000)
44. S. Ishibashi, Y. Higuchi, Y. Ota, J. Vac. Sci. Technol. A **8**, 1399 (1990)
45. K. Zhang, F. Zhu, C.H.A. Huan, A.T.S. Wee, J. Appl. Phys. **86**, 974 (1999)
46. J.-L. Chung, J.-C. Chen, C.-J. Tseng, Appl. Surf. Sci. **255**, 2494 (2008)
47. M. Takashiri, K. Takano, J. Hamada, Thin Solid Films **664**, 100 (2018)
48. M. Takashiri, T. Shirakawa, K. Miyazaki, H. Tsukamoto, Trans. Jpn. Soc. Mech. Eng. Ser. **72**, 1793 (2006)
49. G. Yuan, Y. Li, N. Bao, J. Miao, C. Ge, Y. Wang, Mater. Chem. Phys. **143**, 587 (2014)
50. J.W.G. Bos, H.W. Zandbergen, M.-H. Lee, N.P. Ong, R.J. Cava, Phys. Rev. B **75**, 195203 (2007)
51. Y. Horio, A. Inoue, Mater. Trans. JIM **47**, 1412 (2006)
52. L.D. Zhao, B.-P. Zhang, W.S. Liu, H.L. Zhang, J.-F. Li, J. Alloys Compd. **467**, 91 (2009)
53. Q. Lognone, F. Gascoin, J. Alloys Compd. **610**, 1 (2014)
54. M. Takashiri, T. Shirakawa, K. Miyazaki, H. Tsukamoto, J. Alloys Compd. **441**, 246 (2007)
55. H. Huang, W. Luan, S. Tu, Thin Solid Films **517**, 3731 (2009)
56. M. Takahashi, Y. Katou, K. Nagata, S. Furuta, Thin Solid Films **240**, 70 (1994)
57. N. Hatsuta, D. Takemori, M. Takashiri, J. Alloys Compd. **685**, 147 (2016)
58. M. Takashiri, T. Makioka, H. Yamamuro, J. Alloys Compd. **764**, 802 (2018)
59. S. Michel, S. Diliberto, N. Stein, B. Bolle, C. Boulanger, J. Solid State Electrochem. **12**, 95 (2008)
60. C. Schumacher, K.G. Reinsberg, R. Rostek, L. Akinsinde, S. Baessler, S. Zastrow, G. Rampelberg, P. Woias, C. Detavernier, J.A.C. Broekaert, J. Bachann, K. Nielsch, Adv. Energy Mater. **3**, 95 (2013)
61. B. Vermeersch, J.-H. Bahk, J. Christofferson, A. Shakouri, J. Alloys Compd. **582**, 177 (2014)
62. C. Navone, M. Soulier, M. Plissonnier, A.L. Seiler, J. Electron. Mater. **39**, 1755 (2010)
63. H.Q. Liu, X.B. Zhao, T.J. Zhu, Y. Song, F.P. Wang, Curr. Appl. Phys. **9**, 409 (2009)
64. M. Zebarjadi, K. Esfarjani, A. Shakouri, Z. Bian, J.-H. Bahk, G. Zeng, J. Bowers, H. Lu, J. Zide, A. Gossard, J. Electron. Mater. **39**, 1755 (2010)
65. Q. Zhang, X. Ai, L. Wang, Y. Chang, W. Luo, W. Jiang, L. Chen, Adv. Funct. Mater. **25**, 966 (2015)
66. Y. Hosokawa, K. Wada, M. Tanaka, K. Tomita, M. Takashiri, Jpn. J. Appl. Phys. **57**, 02CC02 (2018)
67. M. Toprak, Y. Zhang, M. Muhammed, Mater. Lett. **57**, 3976 (2003)
68. F.-J. Fan, Y.-X. Wang, X.-J. Liu, L. Wu, S.-H. Yu, Adv. Mater. **24**, 6158 (2011)
69. Y. Zhao, J.S. Dyck, B.M. Hernandez, C. Burda, J. Phys. Chem. C **114**, 11607 (2010)

70. J.S. Son, M.K. Choi, M.-K. Han, K. Park, J.-Y. Kim, S.J. Lim, M. Oh, Y. Kuk, C. Park, S.-J. Kim, T. Hyeon, Nano Lett. **12**, 640 (2012)
71. H.T. Zhang, X.G. Luo, C.H. Wang, Y.M. Xiong, S.Y. Li, X.H. Chen, J. Cryst. Growth **265**, 558 (2004)
72. C. Chen, Z. Ding, Q. Tan, H. Qi, Y. He, Powder Technol. **257**, 83 (2014)
73. A.I. Hochbaum, R. Chen, R.D. Delgado, W. Liang, E.C. Garnett, M. Najarian, A. Majumdar, P. Yang, Nature **451**, 163 (2008)
74. M. Takashiri, T. Shirakawa, K. Miyazaki, H. Tsukamoto, Int. J. Transp. Phenom. **9**, 261 (2007)
75. M. Mizoshiri, M. Mikami, K. Ozaki, Jpn. J. Appl. Phys. **52**, 06GL07 (2013)
76. P. Fan, Z.-H. Zheng, Z.-K. Cai, T.-B. Chen, P.-J. Liu, X.-M. Cai, D.-P. Zhang, G.-X. Liang, J.-T. Luo, Appl. Phys. Lett. **102**, 033904 (2013)

Chapter 2
Wearable Thermoelectric Devices

Seungwoo Han

2.1 Introduction

Wearable thermoelectric devices are of particular interest as alternative power sources for wearable devices, which are becoming more and more popular. Because wearable devices are worn on the human body, it is possible to generate electricity with thermoelectric devices using body heat, without the need for other heat sources.

The most significant issue for a wearable thermoelectric device is flexibility; it must maintain its performance even when bent during use. Unlike conventional thermoelectric devices, the materials, fabrication processes, and device design must be modified to impart flexibility to the devices. Existing rigid ceramic substrates cannot be used, nor can bulk thermoelectric materials. The joint between the electrodes and the thermoelectric legs in a thermoelectric device reduces its flexibility. Hence, the structure of the thermoelectric device may need to be changed to provide the necessary flexibility.

Wearable thermoelectric devices are subjected to cyclic loads during continuous use. Repetitive loading generates cracks within the thermoelectric device, thereby increasing the internal resistance of the device and eventually causing it to fail. Therefore, to develop and commercialize a wearable thermoelectric device, it is essential to establish a method for evaluating the reliability of the thermoelectric element under cyclic loading.

Herein, the current state of wearable thermoelectric device research is reviewed in terms of materials, devices, and applications. A method for evaluating the reliability of a wearable thermoelectric device is presented.

S. Han (✉)
Department of Nano Applied Mechanics, Korea Institute of Machinery and Materials, Daejeon, South Korea
e-mail: swhan@kimm.re.kr

© Springer Nature Switzerland AG 2019
P. Mele et al. (eds.), *Thermoelectric Thin Films*,
https://doi.org/10.1007/978-3-030-20043-5_2

2.2 Materials

When a wearable thermoelectric device is used, the bending stress generated increases with increasing device thickness; a more detailed description of this is given in Sect. 5. Therefore, it is advantageous to apply films of thermoelectric materials to wearable thermoelectric devices, rather than using bulk thermoelectric materials.

We et al. [1] synthesized a thermoelectric paste and used it to fabricate flexible thermoelectric devices. The paste was composed of metal powders (75 wt%; Bi, Sb, and Te), glass powder (2.4 wt%), binder (0.2 wt%), and solvent (22.4 wt%). The high-purity (99.5%) metal powders consisted of particles smaller than 5 mm. The Bi:Te powder mixing ratio was 35:65 and the Sb:Te ratio was 40:60 (ratios expressed as atomic percentages). The pastes were thoroughly mixed for 24 h using a ball mill. The glass powder improved adhesion between the substrate and the paste. The electrical conductivities of the formed Sb_2Te_3 and Bi_2Te_3 films were 3.41×10^4 and 0.73×10^4 S/m, respectively, and their Seebeck coefficients were 92.6 and $-137.8\ \mu V/K$, respectively.

Kim et al. [2] made printable thermoelectric inks. Powdered $Bi_{0.5}Sb_{1.5}Te_3$ and $Bi_2Se_{0.3}Te_{2.7}$ were suspended in an adhesive binder solution. The suspension composed of 78–82 wt% powder and 18–22 wt% binder was mixed with a vortex mixer. A dispenser was used to fill the holes in the fabric with the thermoelectric ink. The viscous printable ink was injected using a syringe. The dispensed ink was dried at room temperature for 24 h, followed by curing at 100 °C in a vacuum oven for 2 h. The synthesized materials were in the form of a uniform and dense matrix. The measured Seebeck coefficients of the formed $Bi_{0.5}Sb_{1.5}Te_3$ and $Bi_2Se_{0.3}Te_{2.7}$ were 0.53 and -0.19 mV/K, respectively.

Francioso et al. [3] deposited thermoelectric thin films on a 50-μm-thick Kapton substrate using radio-frequency magnetron cosputtering (Fig. 2.1). High-purity bismuth (99.999%), antimony (99.555%), and tellurium (99.555%) four-inch-diameter

Fig. 2.1 Scanning electron microscopy images of deposited bismuth telluride (left) and antimony telluride (right) thin films. Reproduced with permission from Francioso et al. [3]

targets were used for the deposition of n- and p-type films. The total argon flow rate during deposition was 25 sccm and the chamber pressure was 7.9×10^{-3} mbar. The deposited p- and n-type thin films were 500-nm thick. Subsequent thermal treatment of the samples was performed in a conventional tubular oven under flowing nitrogen gas (5 L/min at 2 bar pressure) at three different temperatures (175 °C, 225 °C, and 275 °C). The mobilities increased with increasing grain size and reached maximum values of 15.10 and 38.37 cm^2/Vs for the p- and n-type materials, respectively.

Researchers have thus achieved flexibility in wearable thermoelectric devices by using thermoelectric thin-film materials, thermoelectric pastes, and thermoelectric inks. In addition to securing improved flexibility and increasing the thermoelectric performance of the material, other techniques such as vacuum deposition, heat treatment, and chemical synthesis had to be developed.

2.3   Device Design and Fabrication

A conventional thermoelectric device made with bulk material is a vertical device in which electric current and heat flow are perpendicular to each other. Vertical thermoelectric devices have the advantage of improving power generation with increasing temperature difference (ΔT) between the upper and lower plates of the device (see Eq. 2.1). However, it is difficult to bond rigid thermoelectric legs to electrodes. This issue must be solved to make commercially viable vertical wearable thermoelectric devices. On the other hand, a planar thermoelectric device in which the electric current and heat flow are both in the horizontal direction is much easier to manufacture and a flexible structure can be readily realized. However, there is a disadvantage in that the high- and low-temperature parts of the device are located close to the heat source, thereby reducing the temperature difference.

The following section provides some representative examples of how research has enabled the development of flexible thermoelectric devices.

$$\text{Power} \propto \Delta T^2 \tag{2.1}$$

To achieve wearable functionality in thermoelectric devices, researchers have proposed replacing the rigid substrate with a flexible one, or using a fabric as a substrate. A new type of device structure was also developed which removes the substrate and supports the device using a flexible, low-thermal-conductivity filler, such as a polymer.

Lu et al. [4] described a wearable vertical thermoelectric power generator based on commercially available silk fabrics. Figure 2.2 shows the fabrication process. A 4×8 cm^2 section of silk fabric coated with a layer of polyvinyl alcohol was pricked with a needle to generate holes at designated locations, to ensure good contact on both the sides of the material. The power generator was composed of 12 thermocouples and thermoelectric material columns of ca. 300-μm thick and 4-mm diameter. The generator converted thermal energy into electricity for temperature

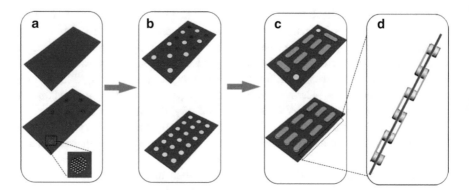

Fig. 2.2 Schematic illustration of the fabrication process of a silk-based thermoelectric power generator. Reproduced with permission from Lu et al. [4]. (**a**) Pretreatment of the silk fabric (**b**) deposition of nanomaterials on the silk fabric (**c**) connection of p-type and n-type columns with silver foils (**d**) side view of the TE power generator

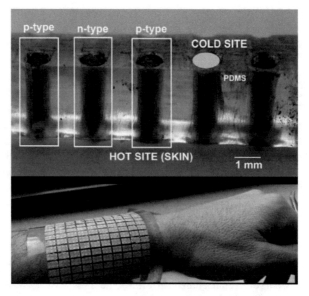

Fig. 2.3 Wearable thermoelectric device containing a polydimethylsiloxane filler. Reproduced with permission from Francioso et al. [5]

differences ranging from 5 to 35 K. The maximum voltage and power outputs were ca. 10 mV and 15 nW, respectively.

Francioso et al. [5] reported a heat-sink-free flexible thermoelectric generator (TEG; Fig. 2.3) fabricated using low-cost screen-printing technology. The designed thermopile consisted of 450 thermocouples prepared by blade coating p- and n-type materials into 2-mm-tall vertical cavities of prepatterned polydimethylsiloxane (PDMS) of total area 98 × 98 mm². The distinguishing feature of this device was

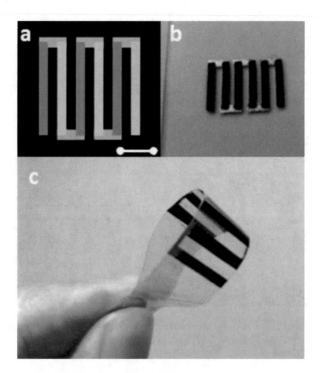

Fig. 2.4 Planar thermoelectric device. Reproduced with permission from Lu et al. [6]. (**a**) Printed device with 3 pairs of p-n legs (**b**) whole printed device on glass (**c**) flexible device on polyimide

the absence of substrate; the PDMS was used as a filler to support the device structure. A thermoelectric power generator including 45 thermocouples and having a footprint area of about 2.2×10^{-3} m^2 generated approximately 27 nW of power at a temperature difference of 10 K.

Lu et al. [6] developed a planar flexible thermoelectric device on polyimide substrates by inkjet printing using a 30-pL piezoelectric printhead (Fig. 2.4). The device was first prepared by printing the silver electrodes followed by sintering at 250 °C for 30 min. In the second step, p- and n-type legs were printed on top of the printed silver electrodes. The thickness of the printed parts of the device could be adjusted by simply repeating the printing process to form as many layers as required. The P and N inks were printed in 150 layers. This research demonstrated the potential of using inkjet printing to fabricate flexible TEGs that are suitable for various applications, such as for harvesting body heat as a wearable power supply.

Kim et al. [2] used a polymer-based fabric as the substrate because of its softness, flexibility, and light weight. A conductive thread was used to make the electrical connections and the fabrics ensured that human wearers could move comfortably. The conductive thread was sewn on the fabric. The prepared printable thermoelectric ink was then dispensed into the windows of the fabric. Figure 2.5 shows the high flexibility and durability of the TEG. After several bending tests, there was no

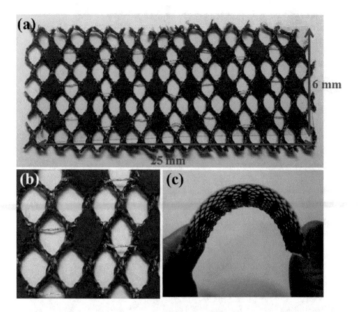

Fig. 2.5 Wearable fabric-based thermoelectric device. Reproduced with permission from Kim et al. [2]. (**a**) the printed 12-couple TEG (**b**) the disenser-printed thermocouples (**c**) bending the wearable TEG by hand

further change in the electrical properties of the TEG. Such testing confirmed that the TEG was sufficiently strong for flexible applications. The total internal resistance of the 12 thermocouples was ca. 902 Ω.

2.4   Applications

Wearable thermoelectric devices are designed and manufactured to be worn by human beings. This section reviews the published research in the context of application areas.

Lu et al. [4] monitored the voltage outputs of an arm-attached generator before and after 30 min of walking at ca. 4 km/h (Fig. 2.6). A significant increase in the voltage output was observed when the device was attached to the arm; 5 to 35 s after the device was attached, the voltage decreased continuously and finally stabilized. After exercise, the voltage output was higher than that for a stationary wearer during the time interval of 5–30 s. The results indicate that the device could effectively harvest waste heat from the human body.

Kim et al. [2] demonstrated that a thermoelectric device can be integrated in the fabric of a shirt (Fig. 2.7). The fabricated device was worn on the wearer's chest because the chest is a large area and is the warmest part of the body. The

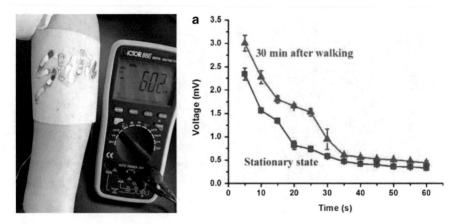

Fig. 2.6 An arm-attached thermoelectric generator. Reproduced with permission from Lu et al. [4]

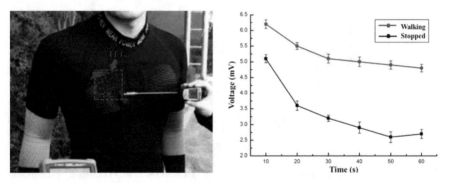

Fig. 2.7 Thermoelectric device applied to a T-shirt. Reproduced with permission from Kim et al. [2]

measured skin temperature was about 32 °C and the ambient temperature was 25 °C. The voltage was highest at first contact of the generator with the skin. After that, the voltage remained at a constant level once thermal equilibrium was achieved. A higher voltage was maintained while walking (at 1 m/s) than in the stationary state. The measured voltages were ca. 2.7 mV for the stationary state and 4.8 mV while walking.

Hyland et al. [7] compared the power-generating performance of thermoelectric devices as a function of location on the human body, i.e., wrist, forearm, or chest, and also when applied to a T-shirt (Fig. 2.8). The highest power produced, ca. 20 μW/cm², was found with the upper arm model when the air speed was fastest, at ca. 1.4 m/s. The least efficient model was the T-shirt model, which consistently produced the lowest power but nevertheless showed reasonable power in the range of 2–8 μW/cm² at similar air velocities. The best power production resulted from the upper arm movement because of the variability of the air flow and good skin contact.

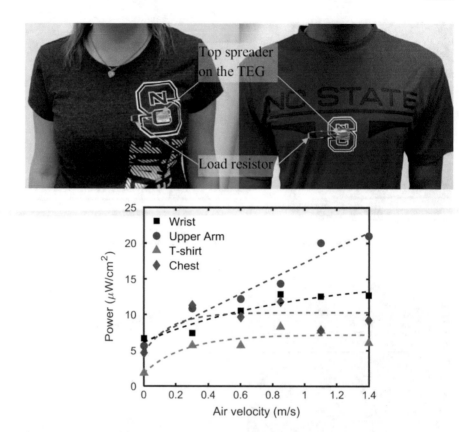

Fig. 2.8 Power output of a thermoelectric device applied to a T-shirt and at different locations on the human body

A potential application of this research is powering a wearable electrocardiogram sensor via a thermoelectric device on the upper arm.

The researchers clearly succeeded in generating electricity by applying a wearable thermoelectric device to different parts of the human body, i.e., wrist, arm, and chest. Maintaining the generated power will increase the utility of wearable thermoelectric devices. To accomplish this, it is necessary to solve the problem of continuous cooling of the side of the thermoelectric device opposite that in contact with the human body.

2.5   Reliability

Wearable thermoelectric devices require power supply systems that can withstand the repetitive bending that is unavoidable in their practical applications. When a wearable thermoelectric device is worn by a user, different bending loads are applied

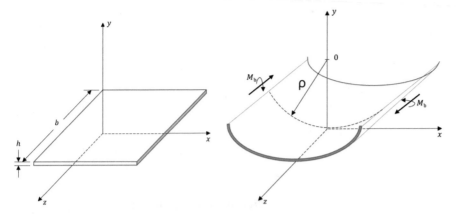

Fig. 2.9 Schematic diagram of the bending load on a wearable thermoelectric device

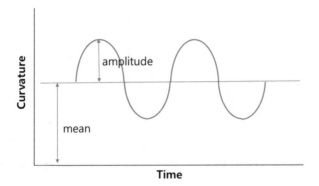

Fig. 2.10 Cyclic loading of a device

to the device. These loads depend on the curvature of the object in contact. Consider a wearable thermoelectric device having thickness h, length b, and being made of a uniform internal material (Fig. 2.9). When a moment M in the longitudinal direction is applied to the thermoelectric device and the device is bent by the radius of curvature ρ, the moment M is expressed by Eq. (2.2) and the radius of curvature ρ is given by Eq. (2.3) [8]. The bending stress σ inside the thermoelectric device is given by Eq. (2.4), where y is the distance from the neutral axis where the bending stress is zero [8]. As y increases, the bending stress acting on the thermoelectric elements increases, which results in the maximum stress σ_{max} at the surface of the device where tension occurs, and the minimum stress σ_{min} at the surface where compression occurs.

A wearable thermoelectric device is subjected to cyclic bending (Fig. 2.10) with repeating wearing and non-wearing. The cyclic bending loads generate fatigue; cracks form in the thermoelectric device, which ultimately fails. The curvature–fatigue life relationship (Fig. 2.11, where the curvature is defined as the reciprocal

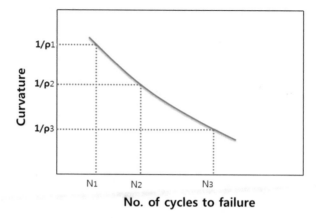

Fig. 2.11 Curvature–fatigue life relationship

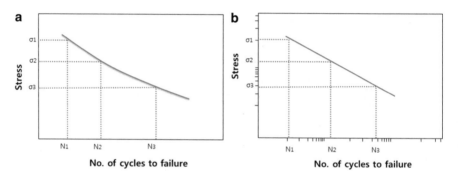

Fig. 2.12 Stress–fatigue life relationship. (**a**) Normal scale and (**b**) log–log scale

of the radius of curvature ρ) can be developed by measuring the resistance of a thermoelectric device while applying a cyclic bending load to it.

The applied bending stress for a given curvature generated by bending the thermoelectric element can be calculated by finite element analysis. The results are then used to obtain the stress–fatigue life relationship (Fig. 2.12) by converting the curvature to the corresponding stress. Fitting the log–log curve in Fig. 2.12 provides the stress–fatigue life relationship of Eq. (2.5), where C and b are the exponents of the fitting equation [9]. If we know the curvature of the surface to which the wearable thermoelectric device is applied, Eq. (2.5) can be used to predict the number of times the thermoelectric device can be worn.

Establishment of this reliability evaluation procedure could provide a basis for commercializing wearable thermoelectric devices.

$$M = -\frac{Eh^3 b}{12\left(1 - v^2\right)\rho} \tag{2.2}$$

$$\frac{1}{\rho} = \frac{12\left(1 - v^2\right)M}{Eh^3b} \tag{2.3}$$

$$\sigma = -\frac{Ey}{1 - v^2}\left(\frac{1}{p}\right) = \frac{12My}{h^3b} \tag{2.4}$$

$$\sigma = CN^{-b} \tag{2.5}$$

2.6 Summary

Research to date on wearable thermoelectric devices has explored many different materials and device technologies for a wide range of applications.

The compositions of pastes, inks, and thin films have been studied extensively for thermoelectric materials. Fabrics, polymeric fillers, flexible substrates, and planar thermoelectric devices have been developed. Wearable thermoelectric devices have been applied in fields that are not suitable for conventional bulk-type thermoelectric devices, e.g., for power generation using body temperature (wrist-, upper arm-, and chest-contact), as wearable sensors, and in a T-shirt.

The fields of application have greatly expanded with the achievement of wearable functionality in thermoelectric devices. It is anticipated that further developments in thermoelectric technology will increase the variety of applications. However, for commercialization, it is essential to develop a technology for evaluating the reliability of a wearable thermoelectric device against cyclic loading during use.

References

1. J.H. We, S.J. Kim, B.J. Cho, Hybrid composite of screen-printed inorganic thermoelectric film and organic conducting polymer for flexible thermoelectric power generator. Energy **73**, 506–512 (2014)
2. M.-K. Kim, M.-S. Kim, S. Lee, C. Kim, Y.-J. Kim, Wearable thermoelectric generator for harvesting human body heat energy. Smart Mater. Struct. **23**, 105002 (2014a)
3. L. Francioso, C. De Pascali, I. Farella, C. Martucci, P. Cretì, P. Siciliano, A. Perrone, Flexible thermoelectric generator for ambient assisted living wearable biometric sensors. J. Power Sources **196**, 3239–3243 (2011)
4. Z. Lu, H. Zhang, C. Mao, C.M. Li, Silk fabric-based wearable thermoelectric generator for energy harvesting from the human body. Appl. Energy **164**, 57–63 (2016)
5. L. Francioso, C. De Pascali, V. Sglavo, A. Grazioli, M. Masieri, P. Siciliano, Modelling, fabrication and experimental testing of an heat sink free wearable thermoelectric generator. Energy Convers. Manag. **145**, 204–213 (2017)
6. Z. Lu, M. Layani, X. Zhao, L.P. Tan, T. Sun, S. Fan, Q. Yan, S. Magdassi, H.H. Hng, Fabrication of flexible thermoelectric thin film devices by inkjet printing. Small **10**(17), 3551–3554 (2014)

7. M. Hyland, H. Hunter, J. Liu, E. Veety, D. Vashaee, Wearable thermoelectric generators for human body heat harvesting. Appl. Energy **182**, 518–524 (2016)
8. S.H. Crandall, N.C. Dahl, T.J. Lardner, *An Introduction to the Mechanics of Solids*, 2nd edn. (McGraw-Hill Book Company, New York, 1978), pp. P490–P491
9. J.A. Bannantine, J.J. Comer, J.L. Handrock, *Fundamentals of Metal Fatigue Analysis* (Prentice-Hall, Inc, Englewood Cliffs, 1990)

Chapter 3
Theory and Simulations of Lattice Thermal Conduction

Matukumilli V. D. Prasad and Umesh V. Waghmare

3.1 Introduction

Heat flow is ubiquitous in nature and occurs at several scales from the thermal evolution of the terrestrial planets to the frictional dissipation in biological systems. Its implications are far-reaching. In condensed matter systems, controlling the heat flow, especially through innovative conduction mechanisms, has been crucial to advances in various technologies such as thermoelectrics, thermal management and information processing [1]. Reducing thermal conduction, for example, can enhance the thermoelectric efficiency. Understanding of the physics of heat transport in nanomaterials can help in engineering the phonon flow to realize exciting applications like hot-spot cooling and waste heat harvesting [2–5].

Calculations of phonon or lattice thermal conductivity, κ, have been carried out through either classical kinetic theory or atomistic molecular dynamics (MD). For simple systems, MD has proven to predict thermal transport accurately [6], it is challenging to treat complex systems due to the difficulty of obtaining accurate interatomic potentials. However, it is interesting to note that MD captures effects of anharmonicity up to all orders. A more popular approach is within the kinetic theory using linearized Boltzmann–Peierls phonon transport equation (BTE) which describes the propagation of interacting phonons in a solid [7]. For many decades, its exact solution has remained as an intractable problem, and several methods have been developed to solve it within some approximations [8]. Omini and Sparavigna's iterative approach [9], partly due to advances in computational resources, succeeded in obtaining numerically exact solution of the full BTE.

M. V. D. Prasad · U. V. Waghmare (✉)
Theoretical Sciences Unit, School of Advanced Materials (SAMat), Jawaharlal Nehru Centre for Advanced Scientific Research (JNCASR), Bangalore, India
e-mail: matukumilli@jncasr.ac.in; waghmare@jncasr.ac.in

© Springer Nature Switzerland AG 2019
P. Mele et al. (eds.), *Thermoelectric Thin Films*,
https://doi.org/10.1007/978-3-030-20043-5_3

Calculations of κ using BTE in conjunction with *ab initio* lattice dynamics have become feasible recently. As they uncover microscopic mechanisms based on phonon scattering rates and scattering phase space (which are difficult to access through experiments), they play an important role in helping establishing the chemical and physical origins of the observed behaviour of κ. The potential of these *ab initio* schemes based on third-order interatomic force constants (IFC) has been established through studies on several classes of materials: Si, Ge [10], diamond [11–13], GaAs [14], compound semiconductors [15], half-Heusler compounds [16], MgO [17, 18], MgSiO$_3$ [19], lead selenide (PbSe), lead telluride (PbTe), and their alloys [20], magnesium silicide (Mg$_2$Si), magnesium stannide (Mg$_2$Sn), and their alloys [21], La$_3$Cu$_3$X$_4$ (X = P, As, Sb, Bi) compounds [22], monolayer InSe, GaSe, GaS, and alloys [23], and boron based cubic bulk [24] and hexagonal 2D [25] compounds.

This chapter is organized as follows: We broadly discuss the theory of lattice thermal conduction in Sect. 3.2. In Sect. 3.3, we detail the simulation procedure. In Sect. 3.4, we present analysis of thermal conductivity in terms of basic vibrational properties for Al, GaAs, and diamond. The potential of BTE in understanding the impact of rattlers on κ of thermoelectric materials is highlighted in Sect. 3.5. In Sect. 3.6, we cover the κ calculations of the materials that are important to understand the Earth's lower mantle. We discuss the current limitations of the BTE approach in addressing the thermal transport in complex molecular crystals in Sect. 3.7. Finally, we conclude with discussion of the main challenges in extending the phonon BTE approach to the phonon gas limit.

3.2 Theory

3.2.1 BTE Under RTA

Under a small temperature gradient, ∇T, across the sample, the induced heat flux would be $\mathbf{Q} = \sum_\lambda \hbar \omega_\lambda \mathbf{v}_\lambda n_\lambda$, [8] where n_λ is the phonon distribution function of phonon frequency ω_λ, and $\mathbf{v}_\lambda$ is the group velocity in x direction of phonon mode, $\lambda = \{\mathbf{q}, p\}$, where $\mathbf{q}$ is the phonon wave vector and p is the branch or polarization index. Under no other external influence, the steady-state phonon distribution function can be tracked by the BTE [26]:

$$-\mathbf{v}_\lambda \cdot \nabla T \left(\frac{\partial n_\lambda}{\partial T} \right) + \left(\frac{\partial n_\lambda}{\partial t} \right)_{scatt} = 0, \qquad (3.1)$$

which represents a balance between the diffusion due to ∇T and collective effect of all scattering processes. The primary issue in solving Eq. 3.1 arises from the scattering term. A popular approximation in solving the BTE is the relaxation time approximation (RTA) where the distribution function n is written as the sum of the equilibrium distribution function, n^{eq} (Bose–Einstein) and fluctuations, n^{fluc}, about equilibrium. Since $n^{fluc} \ll n^{eq}$ and assuming n^{fluc} is independent of T, we can write: $\frac{\partial n}{\partial T} \approx \frac{\partial n^{eq}}{\partial T}$, and by substituting $-\frac{n^{fluc}}{\tau}$ for the scattering term in BTE (where τ is the relaxation time) we get

$$n_\lambda^{fluc} = -\tau_\lambda \, \mathbf{v}_\lambda \cdot \nabla T \frac{\partial n_\lambda^{eq}}{\partial T} \tag{3.2}$$

Determination of κ using the BTE within RTA is possible once the relaxation times τ's are known. Although τ needs to be computed by other means to estimate κ, RTA facilitates writing of the complex scattering term in Eq. 3.1 in terms of n^{fluc} and τ.

3.2.2 Expression of the Relaxation Time

In general, phonons are scattered by boundaries, defects, isotopes, electrons, and other phonons, and each of these processes involves a different scattering rate. Assuming these scattering processes are independent of each other, Matthiessen rule gives the effective scattering rate as:

$$(\tau_{eff})^{-1} = \sum_i (\tau_i)^{-1} \tag{3.3}$$

In case of phonon–phonon scattering, contribution of four- and higher-order phonon interactions to thermal resistance is often insignificant [27] and the interactions involving three phonons dominate the scatting rates.

Using anharmonic IFCs, three-phonon scattering rates, τ_λ^{-1}, are computed using Fermi's golden rule, from the sum of all possible transition probabilities (Γ) for mode λ with modes λ' and λ'' (for absorption processes $\Gamma^+ : \lambda + \lambda' \rightarrow \lambda''$, and emission processes $\Gamma^- : \lambda \rightarrow \lambda' + \lambda''$):

$$\Gamma_{\lambda\lambda'\lambda''}^{\pm} = \frac{\hbar\pi}{4N} \frac{\left(n_\lambda^{eq} + 1\right)\left(n_{\lambda'}^{eq} + \frac{1}{2} \pm \frac{1}{2}\right) n_{\lambda''}^{eq}}{\omega_\lambda \omega_{\lambda'} \omega_{\lambda''}}$$
$$\times \delta\left(\omega_\lambda \pm \omega_{\lambda'} - \omega_{\lambda''}\right) \times \left|\Phi_{\lambda\lambda'\lambda''}^{\pm}\right|^2 \tag{3.4}$$

where N is the number of unit cells and the energy conservation is maintained by the δ function. The three-phonon matrix element, or the strength of interaction among the three phonons, in Eq. 3.4, $\Phi_{\lambda\lambda'\lambda''}$, is expressed as

$$\Phi^{\pm}_{\lambda\lambda'\lambda''} = \sum_k \sum_{l'k'} \sum_{l''k''} \sum_{\alpha\beta\gamma} \Phi_{\alpha\beta\gamma}\left(0k, l'k', l''k''\right)$$

$$\times \frac{e^{\lambda}_{\alpha k} e^{\pm\lambda'}_{\beta k'} e^{-\lambda''}_{\gamma k''}}{\sqrt{M_k M_{k'} M_{k''}}} e^{iq' R_{l'}} e^{iq'' R_{l''}} \tag{3.5}$$

where $\Phi_{\alpha\beta\gamma}$ is real-space anharmonic IFCs and takes the form as:

$$\Phi_{\alpha\beta\gamma}\left(0k, l'k', l''k''\right) = \frac{\partial^3 E}{\partial u_\alpha(0k)\partial u_\beta(l'k')\partial u_\gamma(l''k'')} \tag{3.6}$$

and M_k is atomic mass of kth atom in Θth unit cell, $e^{\lambda}_{\alpha k}$ is the αth component of eigen vector of phonon λ.

While analysing the anharmonic effects in thermal transport, scattering phase space has been identified recently as an important parameter along with scattering rates and anharmonic IFCs [22, 28, 29], which is the sum of the frequency-containing factors in the expression of three-phonon scattering rates (Eq. 3.4) and is written as:

$$W^{\pm}_\lambda = \sum_{\lambda',\lambda''} \frac{\left(n^{eq}_\lambda + 1\right)\left(n^{eq}_{\lambda'} + \frac{1}{2} \pm \frac{1}{2}\right) n^{eq}_{\lambda''}}{\omega_\lambda \omega_{\lambda'} \omega_{\lambda''}} \times \delta\left(\omega_\lambda \pm \omega_{\lambda'} - \omega_{\lambda''}\right) \tag{3.7}$$

Scattering phase space essentially measures the availability of actual energy and momentum conserving three-phonon scattering processes, assuming the strength of interaction $\Phi_{\lambda\lambda'\lambda''}$ is constant [30].

The scattering rates due to three-phonon processes are in the form:

$$(\tau^{RTA}_\lambda)^{-1} = \frac{1}{N}\left(\sum_{\lambda',\lambda''}^{+} \Gamma^{+}_{\lambda\lambda'\lambda''} + \frac{1}{2}\sum_{\lambda',\lambda''}^{-} \Gamma^{-}_{\lambda\lambda'\lambda''}\right) \tag{3.8}$$

In addition to the three-phonon scattering process, one can also include the contribution of isotopic disorder to scattering rate, as is given by Tamura [31]:

$$\Gamma_{\lambda\lambda'} = \frac{\pi\omega^2}{2} \sum_{i \in u.c.} g(i)|\mathbf{e}^*_\lambda(i) \cdot \mathbf{e}_{\lambda'}(i)|^2 \delta(\omega_\lambda - \omega_{\lambda'}) \tag{3.9}$$

where g is the mass variance parameter: $g_i = \sum_j c^j_i \{(m^j_i - \overline{m}_i)/\overline{m}_i\}^2$, where m^j_i is the mass of isotope j of atom i, c^j_i is its concentration and $\overline{m}_i$ is the average isotope mass.

3.2.3 From RTA and Semi-Empirical Scheme to Full Solution: Self-Consistent Calculation of the Relaxation Times

The three-phonon scattering processes are constrained to obey the energy and quasi-momentum conservation up to a reciprocal lattice vector:

$$\hbar\omega_\lambda \pm \hbar\omega_{\lambda'} = \hbar\omega_{\lambda''}, \quad \hbar\mathbf{q} \pm \hbar\mathbf{q}' = \hbar\mathbf{q}'' + a\hbar\mathbf{G} \tag{3.10}$$

where $\mathbf{G}$ is a reciprocal lattice vector and a is an integer whose value is zero for normal (N) and nonzero for Umklapp (U) scattering processes. The $\pm$ signs indicate the two possible three-phonon processes: absorption and emission. At first sight, it appears that N-processes cannot be thermally resistive as the resultant phonon stays within the first BZ. However, treating them as non-resistive processes led to failure in describing experimental κ trends successfully [8, 11, 18, 26]. This leads to the development of semi-empirical models as extension of the RTA by Callaway and Holland. By highlighting the N-process as an indirect contributor to redistributing the energy among different phonons, Callaway model writes the scattering term in BTE (Eq. 3.1) as:

$$\left(\frac{\partial n}{\partial t}\right)_{scatt} = \frac{n^N - n}{\tau_N} + \frac{-n^{fluc}}{\tau_{eff}}, \tag{3.11}$$

where n^N is the equilibrium distribution of phonons involved in N-processes and τ_N is the relaxation time of the N-processes. This model predicts the trends in experimental κ correctly with a few fitting parameters. Holland developed this model further by treating the longitudinal and transverse phonons separately to predict even the trends in high temperature experimental κ correctly. It is important to note that the fitting parameters and the approximations used in these semi-empirical models often obscure the fine details of phonon scattering mechanisms. Towards a parameter-free, and a self-consistent calculation of phonon relaxation times, Omini and Sparavigna's iterative approach [9] has been commonly used in the recent years. As this iterative approach considers changes in phonon distributions due to phonon–phonon scattering events, a more rigorous treatment of anharmonic scattering became possible. Through several recent studies using this iterative approach, it is now evident that the RTA underestimates κ: by $\sim$50% in diamond [11], by $\sim$50% in MgO [18], and by $\sim$39% in AlSb [15].

The mode-wise relation of fluctuations in distribution to τ written explicitly in (Eq. 3.2) and the approximations in semi-empirical models do not allow estimation of precise τ. Now, aiming at the deviation in the distribution function due to phonon–phonon scattering itself, we can rewrite Eq. 3.2 as:

$$n_\lambda^{fluc} = -\mathbf{F}_\lambda \cdot \nabla T \frac{\partial n_\lambda^{eq}}{\partial T}, \tag{3.12}$$

where $\mathbf{F}_\lambda$ is the generalized mean free path. Unlike $\tau_\lambda^{RTA}\mathbf{v}_\lambda$, it considers the deviation in distribution functions. The resulting linearized BTE is

$$\mathbf{F}_\lambda = \tau_\lambda^{RTA}(\mathbf{v}_\lambda + \Delta_\lambda), \tag{3.13}$$

where Δ_λ (in the units of velocity) indicates the deviation (from RTA solution) in the population of a phonon mode, and $\mathbf{F}_\lambda$ can be regarded as a vectorial *mean free displacement,* whose direction, unlike $\tau_\lambda^{RTA}\mathbf{v}_\lambda$, may deviate from the direction of group velocity of a phonon mode.

The converged Δ_λ can be obtained iteratively:

$$\Delta_\lambda = \frac{1}{N} \sum_{\lambda',\lambda''}^{+} \Gamma_{\lambda\lambda'\lambda''}^{+} (\xi_{\lambda\lambda''} F_{\lambda''} - \xi_{\lambda\lambda'} F_{\lambda'})$$
$$+ \frac{1}{2N} \sum_{\lambda',\lambda''}^{-} \Gamma_{\lambda\lambda'\lambda''}^{-} (\xi_{\lambda\lambda''} F_{\lambda''} + \xi_{\lambda\lambda'} F_{\lambda'}) \tag{3.14}$$

$\xi_{\lambda\lambda'}$ is a shorthand for $\omega_{\lambda'}/\omega_\lambda$. For both the absorption and emission three-phonon processes, these sums are over the phase space of λ' and λ'' ensuring energy and quasi-momentum conservation subject to Eq. 3.10. We can initiate the iterative procedure by setting $F_{\lambda'} = F_{\lambda''} = 0$, i.e. by making the zeroth-order solution equivalent to RTA ($\Delta_\lambda = 0$). The ith iterative solution takes the $(i-1)$th iterative solution $\mathbf{F}_\lambda^{i-1}$, $\Gamma_{\lambda\lambda'\lambda''}^{\pm}$ as input and continues until it converges to arrive at the full solution of BTE in the form of $\mathbf{F}_\lambda$ and n_λ of all phonon modes. In contrast, RTA drastically simplifies BTE by ignoring the effects of phonon–phonon scattering on the distribution functions.

Once the converged n_λ and τ_λ are obtained, the phonon thermal conductivity, κ, at temperature, T, can be calculated as the sum of the contributions of all the phonon modes, λ

$$\kappa_{\alpha\beta} = \frac{1}{N\Omega} \sum_\lambda \frac{\partial n_\lambda}{\partial T} (\hbar\omega_\lambda) v_\lambda^\alpha v_\lambda^\beta \tau_\lambda, \tag{3.15}$$

or implemented in the form:

$$\kappa_{\alpha\beta} = \frac{1}{k_B T^2 N\Omega} \sum_\lambda n_\lambda^{eq} (n_\lambda^{eq} + 1)(\hbar\omega_\lambda)^2 v_\lambda^\alpha F_\lambda^\beta \tag{3.16}$$

where N, Ω are the number of q points sampled in integration over Brillouin zone and unit cell volume, respectively.

We note here that the definition of τ will be compromised beyond the RTA as the deviation from n^{eq} became the fundamental parameter in the full BTE solution. Thus, the nature of final quantities, τ, and MFP, Λ, are no longer the same as they are described within the kinetic theory. However these are embedded in the $\mathbf{F}_\lambda$'s to calculate κ. When the deviation of these τ and Λ from those within the kinetic theory is large, the RTA fails badly.

3.3 Simulation Procedure

We present here a procedure in which harmonic and anharmonic IFCs are obtained using first-principles calculations within density functional theory (DFT) and density functional perturbation theory (DFPT) (e.g. as implemented in Quantum Espresso [32] code). We illustrate it through applications to Al, GaAs, and diamond. In DFT calculations, we employed a generalized gradient approximation (GGA) to exchange correlation energy functional and used norm-conserving pseudo-potentials to treat the interactions between ionic cores and valence electrons, and a plane wave basis truncated with an appropriate energy cut-off in the representation of the Kohn–Sham wave functions. The discontinuity in the occupations numbers of electronic states near the gap or Fermi level was smeared with Fermi–Dirac distribution function with a suitable broadening parameter.

To use Fourier transform for obtaining IFCs at any arbitrary q-point, harmonic IFCs are determined at q-vectors on a mesh using DFPT linear response. In calculation of anharmonic third-order IFCs, we employed finite difference approach and used periodic supercells. ShengBTE's *third-order.py* code is used to generate displaced configurations of $4 \times 4 \times 4$ supercell with up to 5th nearest neighbour interactions and the resulting atomic forces are computed in QE. We need these first-principles harmonic and anharmonic IFCs in the ShengBTE [33] code to solve BTE using iterative method [9].

3.4 Illustration of Computation of Lattice Thermal Conductivity

3.4.1 Aluminium

Extensive studies have been devoted to calculate the κ of various nonmetals which lead to the better understanding of phonon thermal transport in semiconductors and insulators [10, 11]. However, relatively less attention has been received by metals due to dominant electronic contribution to thermal transport in many of the metals. Though, the coupled el–ph heat transfer can be addressed through two-temperature model in metals, which is a phenomenological in nature, it fails in the case of semiconductors due to their heterogeneous nature of el–ph interactions. Very recently, phonon-limited thermalization has been identified in semiconductors using a generalized two-temperature model [34] wherein both electrons and phonons do not remain in distinct thermal equilibria. Such simultaneous treatment of electronic and phononic subsystems could lead to the computation of evolution of nonequilibrium phonon distributions emerged due to electronic excitations and their effects on the further thermal relaxation of electrons.

Aluminium is an interesting example for the study of phonon thermal transport. Since Al is a free-electron-like metal, which can be demonstrated from the quadratic behaviour of its electronic DOS, it is likely that el–ph coupling can significantly affect the phonon thermal transport, κ. Unlike Ni and Pt, the electron–phonon scattering rates, τ_{el-ph}^{-1}, in Al are insignificant in comparison with its phonon–phonon scattering rates, τ_{ph-ph}^{-1}. Wang et al. [35] calculated the thermal conductivity and phonon scattering rates of several metals, including Al. Among the studied metals, they found that τ_{el-ph}^{-1} is negligible in Cu, Ag, Au and Al at room temperature whereas it is found to be significant in Ni and Pt. Jain and McGaughey [36] also studied the thermal transport in Al, Ag and Au by considering both electron–phonon and phonon–phonon scattering.

Phonon dispersion of Al in the high symmetry directions in the BZ calculated in this work is shown in Fig. 3.1. The calculated κ (Fig. 3.2a) exhibits reduction with temperature due to enhanced ph–ph anharmonic scattering as evident in Fig. 3.2b. The BTE is solved on three different grids: $15 \times 15 \times 15$, $20 \times 20 \times 20$ and $30 \times 30 \times 30$ and values of κ differ no more than 3% at low temperatures, which indicates that the $15 \times 15 \times 15$ grid is sufficient to give converged κ values. The calculated estimates of κ are in reasonable agreement with the theoretical estimates of Wang et al. [35].

3.4.2 Diamond

κ of diamond can be higher than 3000 W/m-K, which is the highest among the known materials [37–39]. Such an extremely high κ is possible due to: (1) high

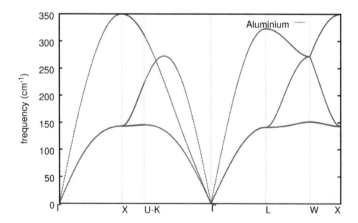

Fig. 3.1 Phonon dispersion of Al

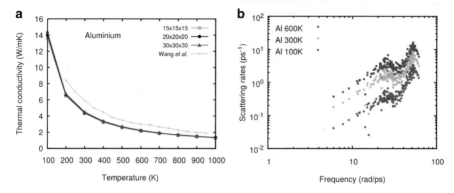

Fig. 3.2 Calculated intrinsic lattice thermal conductivity of Al as a function of temperature from the fully converged solution at three different grids: $15 \times 15 \times 15$ (green squares), $20 \times 20 \times 20$ (black circles) and $30 \times 30 \times 30$ (red triangles), compared with the κ results of Wang et al. [35] (cyan crosses) involving ph–ph interactions (**a**). (**b**) Anharmonic ph–ph scattering causes the $1/T$ behaviour of κ-T relation. Anharmonic phonon scattering rates in Al at three different temperatures: 100 K (red), 300 K (orange) and 600 K (blue) showing stronger scattering at elevated temperatures

bond strength and low atomic mass giving large acoustic velocities and (2) weak ph–ph Umklapp processes. The weak U processes in diamond made it difficult to calculate κ using BTE within RTA. At room temperature, κ estimated with iterative solution is 50% higher value than κ estimated within RTA [11]. In semiconductors, this difference is only $\sim$10% due to the presence of strong U scattering. Such particularly weak U scattering in diamond in comparison with semiconductors like Si and Ge has also been indicated from the results of available three-phonon scattering phase space [30]. For diamond, the three-phonon scattering phase space is surprisingly small in comparison with these semiconductors. Several other examples where RTA underestimates κ include AlSb by 39% [40], MgO by 30% [18].

In Fig. 3.3a we present the calculated κ for diamond by iteratively solving the BTE. These values are in good agreement with the earlier experimental [38, 39] and theoretical [24] results. RTA's under-prediction of κ w.r.t the full solution is also another signature of the correctness of our calculations. From these trends, it is clear that κ from the iterative solution showed 43% higher value than κ obtained within RTA at 200 K. At 500 K, it is only 32% higher.

In comparison with κ of Al (from Fig. 3.2a), diamond exhibits extremely large κ throughout the studied temperature range. Besides, in both the cases, the decreasing trend of κ with temperature is due to the enhanced anharmonic ph–ph scattering. Calculated anharmonic phonon scattering rates of diamond (see Fig. 3.3b) are two orders of magnitude lower than the scattering rates of Al, which explains extremely high κ of diamond.

Several other scenarios have also uncovered through *ab initio* simulations of thermal transport where κ can even be higher than diamond [24]. Studies of Lindsay et al., on κ of a class of boron based cubic compounds (BN, BP, BAs and BSb) above

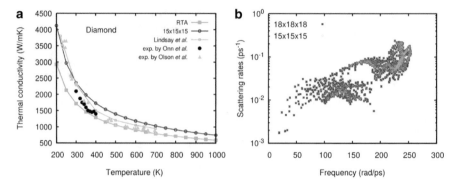

Fig. 3.3 Intrinsic lattice thermal conductivity of diamond as a function of temperature within RTA (green squares) and with fully converged solution (blue circles) at $15 \times 15 \times 15$ grid. Our results are compared with those of experiments by Onn et al. [38] (black circles) and Olson et al. [39] (cyan triangles), and *ab initio* study by Lindsay et al. [24] (orange squares) (**a**). (**b**) Anharmonic phonon scattering rates in diamond computed as a function of frequency obtained on two different q-point grids: $15 \times 15 \times 15$ (orange circles), $18 \times 18 \times 18$ (blue squares)

room temperature showed that BAs surprisingly showed higher κ than diamond. This unusually high κ has been explained, from the phonon dispersion of BAs, as originating from the interplay of: (1) a large frequency gap between acoustic and optic phonons and (2) the bunching of the acoustic branches. This highlights the capabilities of advanced computational techniques for determination of thermal transport properties, and has opened up novel $\omega(q)$ features beyond the conventional criteria for designing high κ materials.

3.4.3 GaAs

With miniaturization of electronic devices, understanding of thermal properties of semiconductors has become crucial. Thermal transport in group IV, III–V and II–VI bulk semiconductors has been investigated using *ab initio* approaches based on the solutions of BTE [30, 40]. GaAs has been extensively studied and widely used in opto- and micro-electronic devices. Hot phonon effects, wherein optical phonons exist in highly nonequilibrium populations in response to the absorbed photo-excited electrons [41, 42], have been attributed to the reduction of carrier cooling rates. Knowledge of phonon relaxation times could help in better understanding of hot phonon effects on the optoelectronic device performance [14]. Semiconductor superlattice structures, where κ can be tuned efficiently, are promising materials especially as thermoelectrics [43]. In Luckyanova et al.'s work [43] based on *ab initio* thermal transport calculations using BTE, the frequency–mean free path relation has been used to explain the coherent nature of phonon transport in GaAs/AlAs superlattice structures.

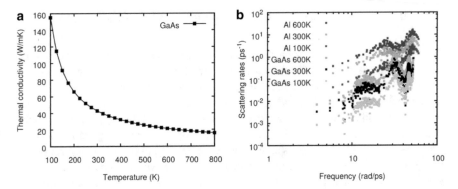

Fig. 3.4 The calculated intrinsic lattice thermal conductivity of GaAs as a function of temperature (**a**). (**b**) Anharmonic phonon scattering rates in GaAs as a function of frequency in comparison with Al at three different temperatures

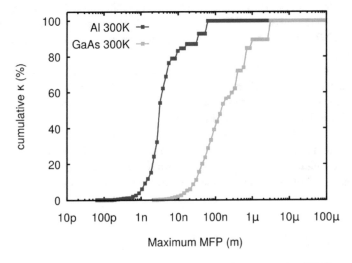

Fig. 3.5 Cumulative thermal conductivity of Al (blue) and GaAs (green) at 300 K. κ values are normalized by their final values (κ_{Al} = 4.12 W/m K and κ_{GaAs} = 42.89 W/m K at 300 K) to bring them on to the same scale for comparison

Figure 3.4a shows the calculated κ values for GaAs. In Fig. 3.4b, we present a comparison of anharmonic scattering rates of GaAs and Al at different temperatures. Overall scattering rates of GaAs are an order of magnitude smaller than those of aluminium.

The normalized cumulative contribution to κ, which indicates a fraction of κ by the phonons of MFP below a certain value, is shown in Fig. 3.5. Comparison between cumulative κ of Al and GaAs provides a clear picture on the thermal transport by phonons with different MFPs. In Al, 90% of κ is contributed by phonons with MFPs less than 10 nm, 90% of κ of GaAs is contributed by its phonons with MFPs less than 1 μm.

3.5 Impact of Rattlers on κ of Thermoelectric Clathrates

A key aspect of designing a high-performance thermoelectric material is to minimize its κ to the theoretical minimum. In general, materials with cage-like crystal
structures (clathrates and skutterudites) exhibit thermal conductivity close to their
theoretical minimum and, initially, this has been attributed to Einstein rattling mode
[44, 45]. In other classes of materials like Zintl type TlInTe$_2$ [46], along with
clathrates and skutterudites [44, 45], the suppression of κ by rattling modes has been
found as an exciting concept, which actually hinted the possibility of nontraditional
phonon scattering scenarios. Intense research efforts have been devoted to identify
efficient routes (such as by intrinsic means) to attain better performing thermoelectric materials by exploiting the material chemistry and bonding nature [47].
Though probing the rattler modes experimentally [48, 49], by means of inelastic X-
ray and inelastic neutron spectroscopies, could provide preliminary understanding
of phonon transport in these guest–host structures, theoretical approaches like *ab
initio* lattice dynamics based BTE, with a detailed information on phonon scattering
dynamics [27], can further provide insights into chemical and physical origins of
ultra-low κ.

Using a simple Born–von Karman (BvK) model, which describes the lattice as
a chain of atoms connected by harmonic springs, Toberer et al. [50] analysed the
effect of mass ratio, number of atoms on κ. With a modified version of this BvK
model where the stiffness of springs altered such that the behaviour of loosely
bound guest atom can be incorporated, they were able to investigate the guest–host
like compounds. Results of analysis of the resonant scattering model are presented
in Fig. 3.6, which emphasize that the information on frequency-dependent group
velocities and scattering rates are vital to unravel the coexisting effects on κ.

3.5.1 Ba$_8$Ga$_{16}$Ge$_{30}$ (Ref. [27])

Here, we highlight the role of anharmonic phonon interaction based calculation of
conductivity in clarifying the impact of a rattler, by discussing results of the study
by Tadano et al. [27] on Ba$_8$Ga$_{16}$Ge$_{30}$. By separately calculating phonon properties
of Ba filled (BGG) and empty Ga$_{16}$Ge$_{30}$ (GG) clathrate structures (see Fig. 3.7),
they observed a tenfold reduction in phonon scattering rates due to ratter dynamics.
Importantly, from the analysis of anharmonic scattering rates (see Fig. 3.8), they
show that the effects of a rattler on the phonon modes are not restricted to a certain
frequency range and thus argued that the phonon scattering in BGG is nonresonant.

Conventionally, ultra-low κ in guest–host structures has been explained by
resonant scattering mechanism, [51, 52] where only a few phonon modes are
affected by the rattler, leading to disordered lattice dynamics, which is popularly
termed as *phonon glass-electron crystal*. Rattler effect is understood by observing:

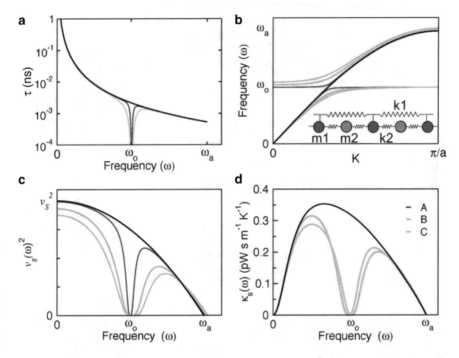

Fig. 3.6 The resonant scattering model targets phonons near ω_0 (**a**). BvK phonon dispersions for a stiff framework (m_1, k_1) and loosely bound guest atoms (m_2, k_2) (**b**). Increased k_2 stiffness results in increased coupling (extent of avoided crossing) between the framework and guest modes. (**c**) The avoided crossing reduces $v_g(\omega)^2$ in the vicinity of ω_0. (**d**) $\kappa_s(\omega)$ for an empty BvK framework, using Umklapp and boundary scattering terms (curve A). Including resonant scattering reduces $\kappa_s(\omega)$ near ω_o (curve B). If instead, the effect of coupling on $v_g(\omega)$ is accounted for, a similar reduction is observed (curve C). Reprinted with permission from Ref. [50]. Copyright 2011 The Royal Society of Chemistry

(1) a shallow potential implying a loose bonding which cause weak restoring forces on the vibrating filler, and (2) lack of phase coherence of the vibrational motion [48].

As implications of crystallographic heterogeneity on anharmonicity are more complex, such *rattling mode* explanation failed in explaining the low thermal conductivity in certain compounds. Other mechanisms such as (1) influence of *flat avoided-crossing bands of filler modes* in phonon scattering, (2) interplay of anharmonicity and scattering phase space also have been identified in explaining low thermal conductivity of skutterudites. In La- and Ce-filled Fe_4Sb_{12} skutterudites [48] it has been observed through neutron spectroscopy experiments and *ab initio* calculations that the filler vibrational modes are *coherently* coupled with the host-lattice dynamics. While this experimental work initiated the counter argument on the resonant nature of the rattler, only the theoretical studies based on BTE have been able to confirm it later by providing a detailed understanding of the phonon behaviour. Tadano et al.'s work [27] is one such example.

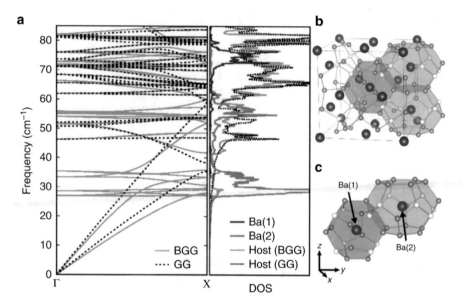

Fig. 3.7 (**a**) Calculated phonon dispersion and projected phonon DOS of BGG (solid lines) and GG (dotted lines). (**b**) Crystal structure of BGG. Ba, Ga and Ge atoms are represented by magenta, white, and blue spheres, respectively. (**c**) Representation of two inequivalent Ba rattlers in dodecahedral (red) and tetrakaidecahedral (blue) polyhedra. Reprinted with permission from Ref. [27]. Copyright 2015 American Physical Society

Phonon dispersion and density of states (DOS) of BGG and GG calculated by Tadano et al. [27] are reproduced in Fig. 3.7a. The rattling modes of Ba(2) atoms can be seen in 28–34 cm^{-1} range along with the avoided crossing of LA modes. From the DOS, Ba(2) induced modes (in blue line) can also be seen at 55 cm^{-1}. These rattling mode frequencies are in good agreement with experiments [53]. Firstly, they have calculated the participation ratio, P_q, for each phonon mode, q, as $P_q = [\sum_\alpha |\mathbf{u}_\alpha(q)|^2]^2 / N_\alpha \sum_\alpha |\mathbf{u}_\alpha(q)|^4$ where $\mathbf{u}_\alpha(q) = M_\alpha^{-1/2} \mathbf{e}_\alpha(q)$, M is atomic mass of atom α and $\mathbf{e}$ is the eigen vector. From the P_q values, they found that the localization of guest atoms is significant in the rattling modes around 28–34 cm^{-1} which is in accordance with the Einstein-rattler picture—a popular notion. However, this participation ratio is an indirect measure of the effect of Ba rattler. To have a direct measure, they used *ab initio* lattice dynamics along with BTE to calculate κ, phonon line width and phonon scattering phase space. The harmonic and cubic IFCs were calculated *ab initio* only for BGG, as the GG structure is fictitious. The trick they have used to generate IFCs for GG from the IFCs of BGG is that they turned off the interactions between the host lattice and guest atoms. At 100 K, calculated κ of GG structure is 20 times greater than that of BGG. Estimated κ of BGG at 100 K is 1.35 W/m-K, which matches well with experimental values (1.5 W/m-K for p-type and 1.9 W/m-K for n-type) [54].

Analysing the scattering rates τ^{-1} of these structures, they obtained important insights: the enhancement of τ^{-1} due to rattler is not limited to any specific frequency region, it is seen in the entire frequency range, profoundly below $120\,\mathrm{cm}^{-1}$. Firstly, this tenfold increase in τ^{-1} explains the 20-fold reduction in κ due to Ba rattler. Secondly, the change of τ^{-1} in entire region is in contradiction with the resonant scattering picture where such effect is localized in frequency domain. Despite these insights from the trends in τ^{-1}, the reason for the increase in τ^{-1} in BGG is still unknown and even the results on three-phonon scattering phase space, $W^{\pm}$ (inset of Fig. 3.8) could not explain it as the increase in $W^{\pm}$ is only by a factor of 2, which is inadequate in explaining the tenfold increase of τ^{-1}. Interestingly, the trends in three-phonon matrix elements (in other words, the strength of interaction among the three phonons), $\Phi_{\lambda\lambda'\lambda''}$, as in Eq. 3.5, appear promising. A significantly higher values of $\Phi_{\lambda\lambda'\lambda''}$ in BGG in comparison with GG have been observed in the frequency ranges precisely corresponding to the frequencies of the localized rattling modes (Fig. 3.7). $\Phi_{\lambda\lambda'\lambda''}$ being a measure of anharmonicity in the system, this increase in $\Phi_{\lambda\lambda'\lambda''}$ is a clear indication of effects of rattler in BGG. This work by Tadano et al. [27] could certainly showcase the enormous potential of the anharmonic lattice dynamics with BTE in understanding the mechanism of thermal transport.

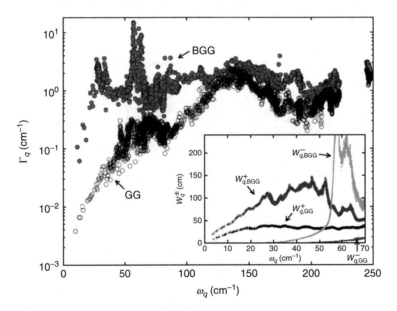

Fig. 3.8 Calculated phonon line width of BGG (filled circles) and GG (open circles) at 300 K. Inset: Energy- and momentum-conserving phase space $W_q^{\pm}$ in the low-energy region at 300 K. Reprinted with permission from Ref. [27]. Copyright 2015 American Physical Society

3.5.2 CoSb₃ *and* BaCo₄Sb₁₂ *(Ref. [29])*

Li and Mingo [29] estimated κ for CoSb₃ and BaCo₄Sb₁₂ skutterudites using *ab initio* lattice dynamics. This is another study where BTE helped in settling the controversial explanation of rattling mechanism. From the comparison between scattering rates of Ba filled skutterudite (BaCo₄Sb₁₂) and empty skutterudite (CoSb₃), they demonstrated that filler causing an increase in scattering rates by a factor of 1.7. Firstly, this enhanced scattering throughout the frequency range ruled out the puzzling resonant nature of rattlers; in the resonant rattler picture such enhancement of scattering is limited to narrow range of frequencies. To uncover the precise role of rattler in τ^{-1}, they calculated the τ^{-1} of BaCo₄Sb₁₂ by *turning off* the interactions between Ba and the host, and found τ^{-1} is still significantly higher than empty skutterudite CoSb₃ (see Fig. 3.9). That indicates that the enhanced scattering in BaCo₄Sb₁₂ is not caused by the presence of filler alone. They also checked the role of third-order IFCs in the enhanced scattering even in the absence of guest–host interactions. For that, they used the CoSb₃'s third-order IFCs in calculating τ^{-1} in BaCo₄Sb₁₂. They did not see much difference in the scattering trends compared to the previous case (see bottom part of Fig. 3.9). Thus, neither the third-order IFCs nor the guest–host interactions cause the enhanced scattering in filled skutterudite. Another possibility is the second-order IFCs which essentially contribute to the scattering rates through frequencies. The three-phonon

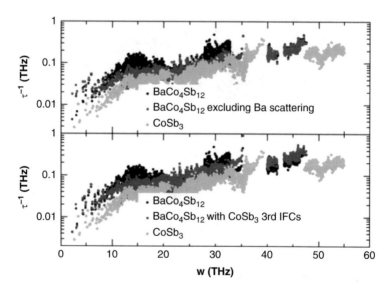

Fig. 3.9 Top: Anharmonic scattering rates of BaCo₄Sb₁₂, compared with those calculated by excluding Ba-related scattering. Bottom: Anharmonic scattering rates of BaCo₄Sb₁₂, compared with those calculated by using the third-order IFCs of CoSb₃. Reprinted with permission from Ref. [29]. Copyright 2014 American Physical Society

scattering phase space, $W^{\pm}$ (see Eq. 3.7) includes frequencies from the transition probabilities equation (see Eq. 3.4). After analysing the $W^{\pm}$ trends, they concluded that the depressed phonon spectrum is the cause for the enhanced scattering in filled skutterudites.

Indeed, such thorough investigation became possible with the valuable and multifaceted information provided by *ab initio* based BTE method.

3.6 Thermal Conductivity of the Earth's Lower Mantle

Thermal structure and evolution of the Earth's interior primarily depends on the heat flow through the Earth's core-mantle boundary, which involves exciting science of thermal boundary layer in terms of heat conduction [55]. As the conduction plays a major role in the heat transport across the core-mantle boundary, an understanding of planetary heat transport requires knowledge of κ as functions of pressure, absolute temperature and chemical composition. This core-mantle thermal structure also has strong correlations with the geomagnetic filed morphology [56]. As the primary constituents of core-mantle boundary layer, polycrystalline aggregate of silicate perovskite ((Fe,Al)-bearing $MgSiO_3$) and ferropericlase ((Mg,Fe)O) have attracted numerous experimental [57–59] and theoretical investigations [17–19, 60, 61]. Earth's core conditions in terms of extreme pressures and temperatures pose technical challenges in measuring κ [57, 58, 62, 63] and thus the extrapolated parameters from ambient conditions have been used in the models to predict behaviour of Earth's interior [64, 65]. In the initial studies, phonon group velocities are approximated as sound velocities and phonon lifetimes at extreme conditions are approximated with lattice thermal expansion and/or Grüneisen parameters which are equilibrium thermodynamic properties.

In the first attempt to calculate κ of MgO, molecular dynamics simulations and Green–Kubo theory have been used [66]. Due to the failure of the interatomic potentials, this study underestimated the κ by an order of magnitude. Later, de Koker [67] used first-principles molecular dynamics along with phonon spectral density approach to arrive at phonon lifetimes, a crucial parameter in estimation of κ. First study that used anharmonic lattice dynamics to calculate κ is by Tang and Dong [17], which marked a significant improvement in modeling κ because of the accuracy in κ it predicted of MgO. Figure 3.10 shows the calculated κ values of MgO as a function of pressure and temperature. They considered a hot and cold geotherm based on experimental constraints [68] to give a direct estimation of κ of MgO at the lower-mantle conditions. These hot and cold geotherms correspond to the whole-mantle convection and partially layered convection, respectively, and provide a reasonable temperature bound for the lower mantle. At the same depth, there is a $\sim$5–10 W/mK difference in κ across the hot and cold geotherms. There is a strong dependence of κ on depth. Though this work based on first-principles calculations succeeded in improving the accuracy of κ estimates of metal oxide materials, it is still far from the realistic values of κ.

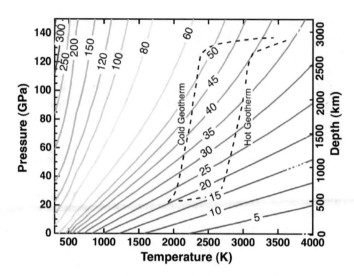

Fig. 3.10 Pressure and temperature contour plot of the lattice thermal conductivity of MgO. Dashed lines are estimated cold and hot geotherms from experiments (Ref. [68]). Reprinted with permission from Ref. [17]. Copyright 2010 National Academy of Sciences

The major concern is the iron enrichment in these mantle minerals which drastically affect the physical and chemical properties of the core-mantle boundary [69, 70]. Recently, there have been studies to understand the effects of iron on κ of Earth's deep mantle and its implications for mantle dynamics [57, 71]. They found even a small percentage of Fe can strongly reduce κ. Similar trends have been found even for larger concentrations of Fe in ferropericlase. Substitution of Al, instead of Fe, in perovskite also let to a significant reduction of κ [57]. There have been no theoretical efforts, using lattice dynamics and BTE, to calculate κ of mantle minerals enriched with small % of Fe or Al.

The other concern, along with the compositional disorder, is the extreme conditions in terms of high pressures and high temperatures, where experimental data is highly scarce and accuracy of theoretical predictions is very poor. For MgO, Dekura and Tsuchiya [18] calculated κ at extreme conditions (up to 150 GPa and 4000 K) using an iterative solution of BTE and compared with RTA results (see Fig. 3.11). They find that RTA underestimates κ of MgO, by about 30%, at both ambient and extreme conditions. Similar to the case of diamond, U phonon scattering is weak in MgO and thus iterative solution of BTE is essential to determine κ accurately.

Iterative solution of BTE can be a potential theoretical tool in improved estimation of κ at Earth's mantle conditions, and can be extended to the other mineral compounds.

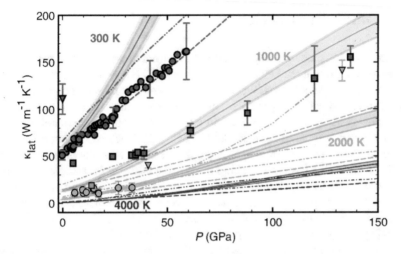

Fig. 3.11 Calculated κ of MgO with the fully converged solutions to the linearized BTE (solid lines) as a function of pressure P at several T from 300 to 4000 K with $1\sigma = 8\%$ confidence bands (shaded regions). The results of previous studies based on *ab initio* ALD calculations with the isotopic correction (dashed double-dotted lines) [17], equilibrium (dash-dotted lines) [72] and nonequilibrium MD (dashed lines) [73], and experiments conducted using single-crystal samples (pink and green circles for 300 and 2000 K, respectively) [62, 63] and polycrystalline samples (pink and orange squares for 300 and 1000 K, respectively) [57, 58] are provided. The results of classical MD simulations (pink and green triangles for 300 and 2000 K, respectively) [74] are also depicted. Reprinted with permission from Ref. [18]. Copyright 2017 American Physical Society

3.7 Molecular Crystals: Limitations of BTE Approach

Phonon behaviour in energetic molecular crystals such as RDX, HMX and TATB is of tantamount importance, especially, to understand the ignition mechanism. For atomic crystals, BTE serves well by providing detailed phonon scattering mechanisms. In molecular crystals such as RDX and TATB energetic materials, it is computationally expensive to deal with large unit cells and it is a long calculation to obtain phonon dispersion. Thus the task of obtaining scattering rates and κ of molecular crystals by solving BTE is computationally not feasible. However, there have been attempts, using other theoretical approaches, to understand the role of phonon scattering in the process of ignition of energetic molecular crystals. In order to calculate κ for energetic material, Long and Chen [75] defined *phonon stress* from individual phonon momenta and obtained phonon–phonon interactions. Based on this, they identified key vibrational modes relevant to heat exchange processes.

Even for atomic crystals, as the crystal complexity increases, inherent anharmonicity starts to dominate and it manifests as weakly dispersed phonon branches, leading to localization of heat.

3.8 Conclusions and Outlook

We have presented computational studies of thermal transport behaviour of several materials to highlight the key features of phonon BTE approach based on *ab initio* lattice dynamics calculations. We started with a detailed presentation of the theory and its developments, and presented our theoretical results for thermal conduction in simple crystals like Al, diamond and GaAs. Moving to the thermoelectrics, we have reviewed how the theoretical studies using BTE helped in demystifying the nature of rattler mechanism—one of the intricate mechanisms of ultra-low κ in thermoelectrics. We have also highlighted the role of theoretical studies using BTE in accurately predicting thermal conductivity of materials under extreme conditions with examples of prominent minerals of Earth's lower mantle. As demonstrated here, phonon BTE along with first-principles lattice dynamics is an exceptional tool, which is yet to be utilized in many other areas and materials.

Nevertheless, extending the phonon BTE beyond the phonon gas limit would be an important step forward. The theoretical foundation of the phonon BTE is routed in the phonon gas model [1, 26], where vibrational modes are weakly interacting such that populations of each phonon follow the single-particle Bose–Einstein distribution at equilibrium. This picture is not true in glassy and amorphous systems. Recently, two-channel model has been proposed [76], where an extra component from the uncorrelated oscillations of vibrational energy was found to be necessary to explain the deviation of phonon BTE estimate of κ from experiments (at high temperatures) in crystalline Tl_3VSe_4. Beyond the phonon gas limit, *vibration Fokker–Planck equation theory* has been proposed to describe the stochastic fluctuation and relaxation processes of lattice vibrations at a wide range of conditions [77]. Such a theory makes multiple-mode correlation functions accessible and has the potential to possibly change the paradigm of thermal physics in the near future.

Acknowledgements UVW thanks for the support from a J. C. Bose National Fellowship of the Department of Science and Technology of Government of India, a Sheikh Saqr Fellowship and IKST-Bangalore.

References

1. G. Chen, *Nanoscale Energy Transfer and Conversion* (Oxford University Press, Oxford, 2005)
2. C. Melnick, M. Kaviany, Phonovoltaic. I. Harvesting hot optical phonons in a nanoscale $p-n$ junction. Phys. Rev. B **93**, 094302 (2016). https://link.aps.org/doi/10.1103/PhysRevB.93.094302
3. C. Melnick, M. Kaviany, Phonovoltaic. II. Tuning band gap to optical phonon in graphite. Phys. Rev. B **93**, 125203 (2016). https://link.aps.org/doi/10.1103/PhysRevB.93.125203
4. C. Melnick, M. Kaviany, Phonovoltaic. III. Electron-phonon coupling and figure of merit of graphene:bn. Phys. Rev. B **94**, 245412 (2016). https://link.aps.org/doi/10.1103/PhysRevB.94.245412

5. A. Seif, W. DeGottardi, K. Esfarjani, M. Hafezi, Thermal management and non-reciprocal control of phonon flow via optomechanics. Nat. Commun. **9**(1), 1207 (2018). ISSN 2041-1723. https://doi.org/10.1038/s41467-018-03624-y
6. B.-L. Huang, M. Kaviany, Ab initio and molecular dynamics predictions for electron and phonon transport in bismuth telluride. Phys. Rev. B. **77**, 125209 (2008). https://link.aps.org/doi/10.1103/PhysRevB.77.125209
7. R. Peierls, Zur kinetischen theorie der wärmeleitung in kristallen. Ann. Phys. **395**(8), 1055–1101 (1929). ISSN 1521-3889. http://dx.doi.org/10.1002/andp.19293950803
8. J.M. Ziman, *Electrons and Phonons: The Theory of Transport Phenomena in Solids* (Oxford University Press, London, 1960)
9. M. Omini, A. Sparavigna, An iterative approach to the phonon boltzmann equation in the theory of thermal conductivity. Phys. B: Condens. Matter. **212**(2), 101–112 (1995). ISSN 0921-4526. https://doi.org/10.1016/0921-4526(95)00016-3. http://www.sciencedirect.com/science/article/pii/0921452695000163
10. A. Ward, D.A. Broido, Intrinsic phonon relaxation times from first-principles studies of the thermal conductivities of Si and Ge. Phys. Rev. B **81**, 085205 (2010). https://link.aps.org/doi/10.1103/PhysRevB.81.085205.
11. A. Ward, D.A. Broido, D.A. Stewart, G. Deinzer, Ab initio theory of the lattice thermal conductivity in diamond. Phys. Rev. B **80**, 125203 (2009). https://link.aps.org/doi/10.1103/PhysRevB.80.125203
12. W. Li, N. Mingo, L. Lindsay, D.A. Broido, D.A. Stewart, N.A. Katcho, Thermal conductivity of diamond nanowires from first principles. Phys. Rev. B **85**, 195436 (2012). https://link.aps.org/doi/10.1103/PhysRevB.85.195436
13. D.A. Broido, L. Lindsay, A. Ward, Thermal conductivity of diamond under extreme pressure: A first-principles study. Phys. Rev. B **86**, 115203 (2012). https://link.aps.org/doi/10.1103/PhysRevB.86.115203
14. T. Luo, J. Garg, J. Shiomi, K. Esfarjani, G. Chen, Gallium arsenide thermal conductivity and optical phonon relaxation times from first-principles calculations.EPL (Europhys. Lett.) **101**(1), 16001 (2013). http://stacks.iop.org/0295-5075/101/i=1/a=16001
15. L. Lindsay, D.A. Broido, T.L. Reinecke, Ab initio thermal transport in compound semiconductors. Phys. Rev. B **87**, 165201 (2013). https://link.aps.org/doi/10.1103/PhysRevB.87.165201
16. J. Shiomi, K. Esfarjani, G. Chen, Thermal conductivity of half-Heusler compounds from first-principles calculations. Phys. Rev. B. **84**, 104302 (2011). https://link.aps.org/doi/10.1103/PhysRevB.84.104302
17. X. Tang, J. Dong, Lattice thermal conductivity of MGO at conditions of earth's interior. Proc. Nat. Acad. Sci. **107**(10), 4539–4543 (2010). ISSN 0027-8424. https://doi.org/10.1073/pnas.0907194107. http://www.pnas.org/content/107/10/4539
18. H. Dekura and T. Tsuchiya, Ab initio lattice thermal conductivity of mgo from a complete solution of the linearized boltzmann transport equation. Phys. Rev. B. **95**, 184303 (2017). https://link.aps.org/doi/10.1103/PhysRevB.95.184303
19. H. Dekura, T. Tsuchiya, J. Tsuchiya, Ab initio lattice thermal conductivity of mgsio$_3$ perovskite as found in earth's lower mantle. Phys. Rev. Lett. **110**, 025904 (2013). https://link.aps.org/doi/10.1103/PhysRevLett.110.025904
20. Z. Tian, J. Garg, K. Esfarjani, T. Shiga, J. Shiomi, G. Chen, Phonon conduction in PbSe, PbTe, and pbte$_{1-x}$se$_x$ from first-principles calculations. Phys. Rev. B **85**, 184303 (2012). https://link.aps.org/doi/10.1103/PhysRevB.85.184303
21. W. Li, L. Lindsay, D.A. Broido, D.A. Stewart, N. Mingo, Thermal conductivity of bulk and nanowire mg$_2$si$_x$sn$_{1-x}$ alloys from first principles. Phys. Rev. B. **86**, 174307 (2012). https://link.aps.org/doi/10.1103/PhysRevB.86.174307
22. T. Pandey, C.A. Polanco, L. Lindsay, D.S. Parker, Lattice thermal transport in La$_3$Cu$_3$X$_4$ compounds (x = P, As, Sb, Bi): Interplay of anharmonicity and scattering phase space. Phys. Rev. B **95**, 224306 (2017). https://link.aps.org/doi/10.1103/PhysRevB.95.224306

23. T. Pandey, D.S. Parker, L. Lindsay, Ab initio phonon thermal transport in monolayer InSe, GaSe, gas, and alloys. Nanotechnology **28**(45), 455706 (2017). http://stacks.iop.org/0957-4484/28/i=45/a=455706

24. L. Lindsay, D.A. Broido, T.L. Reinecke, First-principles determination of ultrahigh thermal conductivity of boron arsenide: a competitor for diamond? Phys. Rev. Lett. **111**, 025901 (2013). https://link.aps.org/doi/10.1103/PhysRevLett.111.025901

25. C. Shi, X. Luo, Characterization of lattice thermal transport in two-dimensional BAs, BP, and BSb: A first-principles study. arXiv:1811.05597v1 (2018)

26. G.P. Srivastava, *The Physics of Phonons* (CRC Press, New York, 1990)

27. T. Tadano, Y. Gohda, S. Tsuneyuki, Impact of rattlers on thermal conductivity of a thermoelectric clathrate: a first-principles study. Phys. Rev. Lett. **114**, 095501 (2015). https://link.aps.org/doi/10.1103/PhysRevLett.114.095501

28. W. Li, N. Mingo, Ultralow lattice thermal conductivity of the fully filled skutterudite $ybfe_4sb_{12}$ due to the flat avoided-crossing filler modes. Phys. Rev. B **91**, 144304 (2015). https://link.aps.org/doi/10.1103/PhysRevB.91.144304

29. W. Li, N. Mingo, Thermal conductivity of fully filled skutterudites: role of the filler. Phys. Rev. B **89**, 184304 (2014). https://link.aps.org/doi/10.1103/PhysRevB.89.184304

30. L. Lindsay, D.A. Broido, Three-phonon phase space and lattice thermal conductivity in semiconductors. J. Phys. Condens. Matter **20**(16), 165209 (2008). http://stacks.iop.org/0953-8984/20/i=16/a=165209

31. S.-I. Tamura, Isotope scattering of dispersive phonons in Ge. Phys. Rev. B **27**, 858–866 (1983). https://link.aps.org/doi/10.1103/PhysRevB.27.858

32. P. Giannozzi, S. Baroni, N. Bonini, M. Calandra, R. Car, C. Cavazzoni, D. Ceresoli, G.L. Chiarotti, M. Cococcioni, I. Dabo, A.D. Corso, S. de Gironcoli, S. Fabris, G. Fratesi, R. Gebauer, U. Gerstmann, C. Gougoussis, A. Kokalj, M. Lazzeri, L. Martin-Samos, N. Marzari, F. Mauri, R. Mazzarello, S. Paolini, A. Pasquarello, L. Paulatto, C. Sbraccia, S. Scandolo, G. Sclauzero, A.P. Seitsonen, A. Smogunov, P. Umari, R.M. Wentzcovitch, Quantum espresso: a modular and open-source software project for quantum simulations of materials. J. Phys. Condens. Matter **21**(39), 395502 (2009). http://stacks.iop.org/0953-8984/21/i=39/a=395502

33. W. Li, J. Carrete, N.A. Katcho, N. Mingo, ShengBTE: a solver of the Boltzmann transport equation for phonons. Comput. Phys. Commun. **185**, 1747–1758 (2014). https://doi.org/10.1016/j.cpc.2014.02.015

34. S. Sadasivam, M.K.Y. Chan, P. Darancet, Theory of thermal relaxation of electrons in semiconductors. Phys. Rev. Lett. **119**, 136602 (2017). https://link.aps.org/doi/10.1103/PhysRevLett.119.136602

35. Y. Wang, Z. Lu, X. Ruan, First principles calculation of lattice thermal conductivity of metals considering phonon-phonon and phonon-electron scattering. J. Appl. Phys. **119**(22), 225109 (2016). https://doi.org/10.1063/1.4953366

36. A. Jain, A.J.H. McGaughey, Thermal transport by phonons and electrons in aluminum, silver, and gold from first principles. Phys. Rev. B **93**, 081206 (2016). https://link.aps.org/doi/10.1103/PhysRevB.93.081206

37. L. Wei, P.K. Kuo, R.L. Thomas, T.R. Anthony, W.F. Banholzer, Thermal conductivity of isotopically modified single crystal diamond. Phys. Rev. Lett. **70**, 3764–3767 (1993). https://link.aps.org/doi/10.1103/PhysRevLett.70.3764.

38. D.G. Onn, A. Witek, Y.Z. Qiu, T.R. Anthony, W.F. Banholzer, Some aspects of the thermal conductivity of isotopically enriched diamond single crystals. Phys. Rev. Lett. **68**, 2806–2809 (1992). https://link.aps.org/doi/10.1103/PhysRevLett.68.2806

39. J.R. Olson, R.O. Pohl, J.W. Vandersande, A. Zoltan, T.R. Anthony, W.F. Banholzer, Thermal conductivity of diamond between 170 and 1200 k and the isotope effect. Phys. Rev. B **47**, 14850–14856 (1993). https://link.aps.org/doi/10.1103/PhysRevB.47.14850

40. L. Lindsay, D.A. Broido, T.L. Reinecke, Ab initio thermal transport in compound semiconductors. Phys. Rev. B **87**, 165201 (2013). https://link.aps.org/doi/10.1103/PhysRevB.87.165201

41. E. G. Lluesma, G. Mendes, C. Arguello, and R. Leite, Very high non-thermal equilibrium population of optical phonons in GaAs. Solid State Commun. **14**(11), 1195–1197 (1974). ISSN 0038-1098. https://doi.org/10.1016/0038-1098(74)90302-0. http://www.sciencedirect.com/science/article/pii/0038109874903020
42. W. Pötz, Hot-phonon effects in bulk GaAs. Phys. Rev. B **36**, 5016–5019 (1987). https://link.aps.org/doi/10.1103/PhysRevB.36.5016
43. M.N. Luckyanova, J. Garg, K. Esfarjani, A. Jandl, M.T. Bulsara, A.J. Schmidt, A.J. Minnich, S. Chen, M.S. Dresselhaus, Z. Ren, E.A. Fitzgerald, G. Chen, Coherent phonon heat conduction in superlattices. Science **338**(6109), 936–939 (2012). ISSN 0036-8075. https://doi.org/10.1126/science.1225549. http://science.sciencemag.org/content/338/6109/936
44. G.S. Nolas, J.L. Cohn, G.A. Slack, S.B. Schujman, Semiconducting Ge clathrates: Promising candidates for thermoelectric applications. Appl. Phys. Lett. **73**(2), 178–180 (1998). https://doi.org/10.1063/1.121747
45. G.S. Nolas, J.L. Cohn, G.A. Slack, Effect of partial void filling on the lattice thermal conductivity of skutterudites. Phys. Rev. B **58**, 164–170 (1998). https://link.aps.org/doi/10.1103/PhysRevB.58.164
46. M.K. Jana, K. Pal, A. Warankar, P. Mandal, U.V. Waghmare, K. Biswas, Intrinsic rattler-induced low thermal conductivity in Zintl type tlinte2. J. Am. Chem. Soc. **139**(12), 4350–4353 (2017). http://dx.doi.org/10.1021/jacs.7b01434. PMID: 28263613
47. W.G. Zeier, A. Zevalkink, Z.M. Gibbs, G. Hautier, M.G. Kanatzidis, G.J. Snyder, Thinking like a chemist: Intuition in thermoelectric materials. Angewandte Chemie International Edition **55** (24), 6826–6841 (2016). ISSN 1521-3773. http://dx.doi.org/10.1002/anie.201508381
48. M.M. Koza, M.R. Johnson, R. Viennois, H. Mutka, L. Girard, D. Ravot, Breakdown of phonon glass paradigm in la- and ce-filled fe4sb12 skutterudites. Nat. Mater. **7**, 805 EP 810 (2008). https://doi.org/10.1038/nmat2260
49. D.J. Voneshen, K. Refson, E. Borissenko, M. Krisch, A. Bosak, A. Piovano, E. Cemal, M. Enderle, M.J. Gutmann, M. Hoesch, M. Roger, L. Gannon, A.T. Boothroyd, S. Uthayaku-mar, D.G. Porter, J.P. Goff, Suppression of thermal conductivity by rattling modes in thermoelectric sodium cobaltate. Nat. Mater. **12**, 1028 EP (2013). https://doi.org/10.1038/nmat3739
50. E.S. Toberer, A. Zevalkink, G.J. Snyder, Phonon engineering through crystal chemistry. J. Mater. Chem. **21**, 15843–15852 (2011). http://dx.doi.org/10.1039/C1JM11754H
51. J.L. Cohn, G.S. Nolas, V. Fessatidis, T.H. Metcalf, G.A. Slack, Glasslike heat conduction in high-mobility crystalline semiconductors. Phys. Rev. Lett. **82**, 779–782 (1999). https://link.aps.org/doi/10.1103/PhysRevLett.82.779
52. J.S. Tse, V.P. Shpakov, V.R. Belosludov, F. Trouw, Y.P. Handa, W. Press, Coupling of localized guest vibrations with the lattice modes in clathrate hydrates. EPL (Europhys. Lett.) **54**(3), 354 (2001). http://stacks.iop.org/0295-5075/54/i=3/a=354
53. Y. Takasu, T. Hasegawa, N. Ogita, M. Udagawa, M.A. Avila, K. Suekuni, I. Ishii, T. Suzuki, T. Takabatake, Dynamical properties of guest ions in the type-i clathrate compounds $X_8ga_{16}ge_{30}$ (x = Eu, Sr, Ba) investigated by raman scattering. Phys. Rev. B **74**, 174303 (2006). https://link.aps.org/doi/10.1103/PhysRevB.74.174303
54. M.A. Avila, K. Suekuni, K. Umeo, H. Fukuoka, S. Yamanaka, T. Takabatake, Glasslike versus crystalline thermal conductivity in carrier-tuned $ba_8ga_{16}X_{30}$ clathrates (x = Ge, Sn). Phys. Rev. B **74**, 125109 (2006). https://link.aps.org/doi/10.1103/PhysRevB.74.125109
55. T. Lay, J. Hernlund, B.A. Buffett, Core-mantle boundary heat flow. Nat. Geosci. **1**, 25 EP (2008). https://doi.org/10.1038/ngeo.2007.44. Review Article
56. D. Gubbins, A.P. Willis, B. Sreenivasan, Correlation of Earth's magnetic field with lower mantle thermal and seismic structure. Phys. Earth Planet. Inter. **162**(3), 256–260 (2007). ISSN 0031-9201. https://doi.org/10.1016/j.pepi.2007.04.014. http://www.sciencedirect.com/science/article/pii/S0031920107000908

57. G.M. Manthilake, N. de Koker, D.J. Frost, C.A. McCammon, Lattice thermal conductivity of lower mantle minerals and heat flux from earth's core. Proc. Nat. Acad. Sci. **108**(44), 17901–17904 (2011). ISSN 0027-8424. https://doi.org/10.1073/pnas.1110594108. http://www.pnas.org/content/108/44/17901

58. A.F. Goncharov, S.S. Lobanov, X. Tan, G.T. Hohensee, D.G. Cahill, J.-F. Lin, S.-M. Thomas, T. Okuchi, N. Tomioka, Experimental study of thermal conductivity at high pressures: Implications for the deep earth's interior. Phys. Earth Planet. Inter. **247**, 11–16 (2015). ISSN 0031-9201. https://doi.org/10.1016/j.pepi.2015.02.004. http://www.sciencedirect.com/science/article/pii/S0031920115000199. Transport Properties of the Earth's Core

59. K. Ohta, T. Yagi, K. Hirose, Y. Ohishi, Thermal conductivity of ferropericlase in the earth's lower mantle. Earth Planet. Sci. Lett. **465**, 29–37 (2017). ISSN 0012-821X. https://doi.org/10.1016/j.epsl.2017.02.030. http://www.sciencedirect.com/science/article/pii/S0012821X17300985

60. A.M. Hofmeister, Mantle values of thermal conductivity and the geotherm from phonon lifetimes. Science **283**(5408), 1699–1706 (1999). ISSN 0036-8075. https://doi.org/10.1126/science.283.5408.1699. http://science.sciencemag.org/content/283/5408/1699

61. H. Aramberri, R. Rurali, J. Íñiguez, Thermal conductivity changes across a structural phase transition: the case of high-pressure silica. Phys. Rev. B. **96**, 195201 (2017). https://link.aps.org/doi/10.1103/PhysRevB.96.195201

62. A.F. Goncharov, P. Beck, V.V. Struzhkin, B.D. Haugen, S.D. Jacobsen, Thermal conductivity of lower-mantle minerals. Phys. Earth Planet. Inter. **174**(1), 24–32 (2009). ISSN 0031-9201. https://doi.org/10.1016/j.pepi.2008.07.033. http://www.sciencedirect.com/science/article/pii/S0031920108001945. Advances in High Pressure Mineral Physics: from Deep Mantle to the Core

63. D.A. Dalton, W.-P. Hsieh, G.T. Hohensee, D.G. Cahill, A.F. Goncharov, Effect of mass disorder on the lattice thermal conductivity of MgO periclase under pressure. Sci. Rep. **3**, 2400 EP (2013). https://doi.org/10.1038/srep02400. Article

64. J.M. Brown, Interpretation of the d" zone at the base of the mantle: dependence on assumed values of thermal conductivity. Geophys. Res. Lett. **13**(13), 1509–1512 (1986). https://doi.org/10.1029/GL013i013p01509. https://agupubs.onlinelibrary.wiley.com/doi/abs/10.1029/GL013i013p01509

65. A.M. Hofmeister, Inference of high thermal transport in the lower mantle from laser-flash experiments and the damped harmonic oscillator model. Phys. Earth Planet. Inter. **170**(3), 201–206 (2008). ISSN 0031-9201. https://doi.org/10.1016/j.pepi.2008.06.034. http://www.sciencedirect.com/science/article/pii/S0031920108002331. Frontiers and Grand Challenges in Mineral Physics of the Deep Mantle

66. R.E. Cohen, Thermal conductivity of MgO at high pressures. Rev. High Pressure Sci. Technol. **7**, 160–162 (1998). https://doi.org/10.4131/jshpreview.7.160

67. N. de Koker, Thermal conductivity of MgO periclase from equilibrium first principles molecular dynamics. Phys. Rev. Lett. **103**, 125902 (2009). https://link.aps.org/doi/10.1103/PhysRevLett.103.125902

68. R. Jeanloz, S. Morris, Temperature distribution in the crust and mantle. Ann. Rev. Earth Planet. Sci. **14**(1), 377–415 (1986). https://doi.org/10.1146/annurev.ea.14.050186.002113

69. J. Badro, Spin transitions in mantle minerals. Ann. Rev. Earth Planet. Sci. **42**(1), 231–248 (2014). https://doi.org/10.1146/annurev-earth-042711-105304

70. J.-F. Lin, S. Speziale, Z. Mao, H. Marquardt, Effects of the electronic spin transitions of iron in lower mantle minerals: implications for deep mantle geophysics and geochemistry. Rev. Geophys. **51**(2), 244–275 (2013). https://agupubs.onlinelibrary.wiley.com/doi/abs/10.1002/rog.20010

71. W.-P. Hsieh, F. Deschamps, T. Okuchi, J.-F. Lin, Effects of iron on the lattice thermal conductivity of earth's deep mantle and implications for mantle dynamics. Proc. Nat. Acad. Sci. **115**(16), 4099–4104 (2018). ISSN 0027-8424. https://doi.org/10.1073/pnas.1718557115. http://www.pnas.org/content/115/16/4099

72. N. de Koker, Thermal conductivity of MgO periclase at high pressure: implications for the d" region. Earth Planet. Sci. Lett. **292**(3), 392–398 (2010). ISSN 0012-821X. https://doi.org/10.1016/j.epsl.2010.02.011.   http://www.sciencedirect.com/science/article/pii/S0012821X10001135
73. S. Stackhouse, L. Stixrude, B.B. Karki, Thermal conductivity of periclase (MgO) from first principles. Phys. Rev. Lett. **104**, 208501 (2010). https://link.aps.org/doi/10.1103/PhysRevLett.104.208501
74. V. Haigis, M. Salanne, S. Jahn, Thermal conductivity of MgO, MgSiO3 perovskite and post-perovskite in the earth's deep mantle. Earth Planet. Sci. Lett. **355–356**, 102–108 (2012). ISSN 0012-821X. https://doi.org/10.1016/j.epsl.2012.09.002. http://www.sciencedirect.com/science/article/pii/S0012821X12004815
75. Y. Long, J. Chen, Theoretical study of the phonon–phonon scattering mechanism and the thermal conductive coefficients for energetic material. Philos. Mag. **97**(28), 2575–2595 (2017). https://doi.org/10.1080/14786435.2017.1343962
76. S. Mukhopadhyay, D.S. Parker, B.C. Sales, A.A. Puretzky, M.A. McGuire, L. Lindsay, Two-channel model for ultralow thermal conductivity of crystalline tl3vse4. Science **360**(6396), 1455–1458 (2018). ISSN 0036-8075. https://doi.org/10.1126/science.aar8072. http://science.sciencemag.org/content/360/6396/1455
77. Y. Zeng, J. Dong, The Fokker-Planck equation for lattice vibration: stochastic dynamics and thermal conductivity. arXiv:1811.08492v1 (2018)

Chapter 4
Fabrication and Thermoelectric Properties of PEDOT Films and Their Composites

Wei Shi, Qin Yao, and Lidong Chen

4.1 Introduction

Polythiophene is electrically polymerized by Tourillon and Garnier et al. in 1982, which exhibits an electrical conductivity (σ) of ~10–100 S/cm [1]. To enhance its performance, stabilized doping states and enlarged conjugation length are desired. Therefore, various derivatives are synthesized. As an example, poly(3-hexylthiophene) [2–6] with higher doping capability and electrical performances are synthesized. However, most of the derivatives are still not stable enough, the doping process is slow and reversible. Some are even dedoped when immersed in solvent such as ethanol. Furthermore, complex synthesis and relatively poor performance limit their practical applications.

Compared to other polythiophene derivatives, PEDOT exhibits much more stabilized doping states. It was initially discovered by Jonas, Heywang et al. of the Bayer Company [7–8]. The effective mesomeric stabilization provided by 3,4-oxygen substitution results in remarkably enhanced doping stability and high σ. The heavily doped bipolaron structure absorbs mostly infrared light, which results in the transparent light blue appearance of the PEDOT films.

Owing to its high electrical performance and good chemical stability, PEDOT has been widely used in various fields such as supercapacitor [9–12], solar cell [13–15], sensors [16–19], and ferroelectric [20]. However, in thermoelectric (TE) material, large Seebeck coefficient (S) is required in addition to high σ. The great structural defect in PEDOT's film has led to limited S and σ. The optimization of the polymer structure and the introduction of nanoparticles are the two main routes to achieve higher TE performance. Great efforts have been done in recent two decades. In this

W. Shi · Q. Yao · L. Chen (✉)
State Key Laboratory of High Performance Ceramics and Superfine Microstructures, Shanghai Institute of Ceramics, Chinese Academy of Sciences, Shanghai, China
e-mail: shiwei@mail.sic.ac.cn; yaoqin@mail.sic.ac.cn; cld@mail.sic.ac.cn

© Springer Nature Switzerland AG 2019
P. Mele et al. (eds.), *Thermoelectric Thin Films*,
https://doi.org/10.1007/978-3-030-20043-5_4

chapter, we will mainly introduce the performance optimization of PEDOT-based TE film material.

4.2 Two Types of PEDOT

PEDOTs are fully doped when polymerized, which can be classified according to their counter-ions. There are basically two types of PEDOTs: the commercially available PEDOT:polystyrenesulfonate (PEDOT:PSS) and chemically produced small-sized anion doped-PEDOT (S-PEDOT). S-PEDOTs were discovered earlier, and are usually obtained via in-situ polymerization or electrochemical polymerization. They are completely insoluble in all solvents, which limit their applications.

PEDOT:PSS was invented a year after the discovery of S-PEDOT [21]. Due to the existence of hydrophilic polyanion PSS, PEDOT:PSS is water-processable. PEDOT:PSS aqueous solutions (mostly ∼1 wt%) are commercially available now and are much more commonly used than S-PEDOT since PEDOT:PSS is mentioned as the polymer matrix in nearly 90% of the reports in PEDOT-based TE material rather than S-PEDOT.

However, either type of PEDOT has its distinctive advantage/disadvantage as TE material. Due to the difference in structure/synthesis, the TE performance of PEDOT:PSS and S-PEDOT is optimized via completely different approaches. And we will discuss separately in the following chapters.

4.3 The TE Performance Optimization of PEDOT:PSS Films

4.3.1 Synthesis of PEDOT:PSS

PEDOT:PSS is synthesized via an oxidation reaction in PSS aqueous solution where EDOT is oxidized by ammonium persulfate (APS) [21]. The by-product and excess oxidant in raw product can be washed off by ion-exchange method. The resulting dark-blue solution can be used directly for film coating. In the early period, only PEDOT:PSS is used in TE material.

4.3.2 TE Properties of Pristine PEDOT:PSS

Jiang et al. first reported the TE performance of PEDOT:PSS (DMSO-treated). The cold-pressed pellet exhibits a ZT value of 1.75×10^{-3} [22]. Though the material exhibits a very low thermal conductivity (κ) of ∼0.1 W/mK, its S (∼12 μV/K) and

Table 4.1 TE properties of various Clevios PEDOT products from the Bayer Company

Products	σ (S/cm)	S (μV/K)	PF (μW/mK^2)
Clevios P (5% DMSO) [25]	81.9	12.6	1.28
Clevios PH500 (5% DMSO) [26]	330	14.6	7.0
Clevios PH500 (5% DMSO) [25]	317	22.5	16.05
Clevios PH510 (5% DMSO) [23]	299	12.6	4.8
Clevios PH750 (5% DMSO) [26]	570	13.5	10.4
Clevios PH1000 (5% DMSO) [25]	945	22.2	46.57
Clevios FET [25]	320	30.3	29.39

σ (~50 S/cm) are not satisfied. The main reason of low performance is the lower density/continuity of powder material compared to that of the film material. Chang et al. first reported the TE performance of PEDOT:PSS film. The coated film with self-made PEDOT:PSS solution exhibits a σ of 250 S/cm and a power factor (PF) of 4.78 μW/mK^2 [23].

The σ of PEDOT:PSS films is further enhanced in the later series of commercial products. For example, Clevios PH-500 stands for a film with a σ close to 500 S/cm. Fan et al. have compared the film properties of Clevios PH-500 with Clevios P and found that the higher electrical performance was probably attributed to the higher molecular weight of the polymer [24]. Zhang et al. have measured the TE properties of various Clevios products. The results showed that 5% DMSO-treated Clevios PH-1000 exhibits a fairly high power factor of 46.57 μW/mK^2 [25]. Table 4.1 summarizes the TE performance of the commercially available PEDOT:PSS products.

4.3.3 Structural Distortion in PEDOT:PSS

However, the bottleneck appears in the synthesis of high performance PEDOT:PSS-based TE material as people found that the σ and S of the these materials reached a limitation of ~1000 S/cm and ~20 μV/K, respectively. Reports show that PEDOT:PSS is not effectively doped. Usually, only a small proportion of sulfonate units in PSS participate in the doping process ($-SO_3^-$) while the rest remain in their acid type ($-SO_3H$) [48]. This unbalanced doping structure results in severe distortion in PEDOT's chain structure (Fig. 4.1a). The carriers are localized in such Fermi glass [27], and the improvement of S is limited. Secondly, segregation forms due to the large and not fully doped PSS segment existing as independent insulating phase in PEDOT and finally results in an island-like structure consist of the PEDOT-rich conductive core area and PSS-rich less-conductive outer layer (Fig. 4.1b). Lang et al. have observed such inhomogeneous PEDOT-PSS binary phase structure via high-angle annular dark field scanning transmission electron microscopy (HAADF-

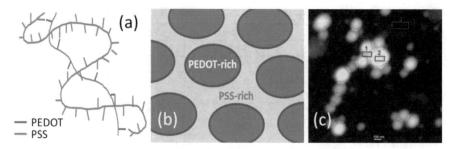

Fig. 4.1 (**a**) Asymmetrical arrangement between PEDOT and PSS. (**b**) The island distribution of conducting PEDOT segments in PEDOT:PSS. (**c**) HAADF-STEM image of PEDOT:PSS film, showing the boundaries between PEDOT-rich segments [28]. Reproduced with permission

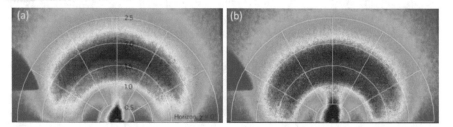

Fig. 4.2 GIWAXS patterns of PEDOT:PSS films (**a**) without and (**b**) with 0.05% DEG addition [27]. Reproduced with permission

STEM) (Fig. 4.1c) [28]. The interfacial boundaries could block the carrier transport and limits the σ of the material. Furthermore, the distribution of PSS is also affected by the synthesizing/coating technique and the in-plane and out-of-plane σ is usually different [29].

4.3.4   Secondary Doping of PEDOT:PSS

As mentioned above, the 5% DMSO addition can significantly enhance the σ of PEDOT:PSS. This process is called secondary doping, in which the solvent with high dielectric constant is used, such as dimethyl sulfoxide (DMSO) [23, 26, 30, 31], ethylene glycol (EG) [32–34], and diethylene glycol (DEG) [27]. Though secondary doping does not result in crystallization of PEDOT, it can improve the structure ordering to a certain extent. Some reports show that when PEDOT is treated with DEG, the ratio of non-variable range hopping (VRH) conduction increased, which indicates that the structural ordering of the material is enhanced. The GIWAXS also shows that an additional peak that refers to the separation of PEDOT and

PSS emerged when PEDOT:PSS is treated with DEG (Fig. 4.2) [27]. These results confirmed that solvent treatment can separate PEDOT and PSS phase and leads to enhancements in σ and structural ordering.

4.3.5   Removal of PSS

In addition to secondary doping, some studies focus on the direct removal of PSS polyanion in PEDOT. Kim et al. immersed the PEDOT:PSS in DMSO or EG for a certain period of time and found that the PPS content in PEDOT is significantly reduced, along with a decrease in film thickness. Both σ and S increased and an amazingly high ZT value of 0.42 is achieved [35]. This method can be regarded as an enhanced version of secondary doping. Except for solvent treatments, Lee et al. reported effective PSS removals by chemical dedoping (hydrazine) [36] and ultra-filtration [37]. By using these methods, the TE performance of PEDOT:PSS film can be maximized and a high power factor of 115.5 μW/mK^2 is achieved. There is also a report that shows a summarized result of chemical dedoping of PEDOT:PSS by various types of reductants [38]. The improvement of S is determined by the dedoping/reducing ability of the reductant.

Recently, acid treatment became a new effective route to remove excess PSS in PEDOT. Through acid solution/vapor treatment (e.g., H$_2$SO$_4$ [39], oxalic acid [40], formic acid [41], HI [42]), the σ of the material is enhanced and the S remains almost unchanged. Such treatments can lower the PEDOT:PSS content and enhance the doping level by anion substitution, which are also proved to be effective in the performance optimization of S-PEDOT such as PEDOT:tosylate (PEDOT:Tos) and PEDOT:FeCl$_4$ [43–45]. Besides, post-treatments using formamide [46], ZnCl$_2$ [47], sorbitol [48–49], and deep eutectic solvent (DES) [50–51] have similar positive effects. The above methods optimizing the structure ordering of PEDOT:PSS are summarized in Table 4.2.

These reports have confirmed the fact that the TE performance of PEDOT:PSS is influenced by the continuity, ordering, and density of the conductive structure. An ideal structure of PEDOT:PSS should have compact and ordered structure with just enough PSS dopant. However, despite its reasonable σ and solvent processability as a successful commercial product, the complete removal of excess PSS and the ordered structure is extremely difficult to realize due to the mismatch between positive charge and SO$_3^-$ anions. PEDOT:PSS exhibits non-crystalline structure and low Seebeck coefficient compared to S-PEDOTs (100% effectively doped) [27]. New strategies are demanded for the further enhancement of PEDOT:PSS matrix. Despite these shortages, PEDOT:PSS has great advantage in the convenient syntheses of nanocomposite materials in which TE performance can be further enhanced (which will be discussed in the later sessions).

Table 4.2 The reported TE properties of PEDOT:PSS (performance optimization through generating a more ordered structure)

Treatments	σ (S/cm)	S (μV/K)	PF (μW/mK^2)
EG/DMSO (secondary doping) [22]	40	12	0.576
DMSO (secondary doping) [23]	298.52	12.65	4.78
DMSO (secondary doping) [25]	945	22.2	46.6
DMSO (secondary doping) [30]	~800	18.98	28.95
DMSO (dilution–filtration) [31]	1399	18.6	48.3
EG (solvent-thermal) [34]	520	16.9	14.9
EG/DMSO (removing PSS) [35]	~880	~74	482
Ultra-filtration (removing PSS) [37]	788	32.5	83.2
Oxalic acid (removing PSS) [40]	823	9.2	6.96
Formic acid (removing PSS) [41]	1900	20.6	80.6
Sorbitol (removing PSS) [48]	722	~10	7.26
Sorbitol (removing PSS) [49]	674.6	13.8	12.8
DES (removing PSS) [51]	424.2	24.4	25.26
ZnCl$_2$-DMF (removing PSS) [47]	~1450	26.1	98.2
H$_2$SO$_4$ vapor (removing PSS) [39]	1167	12.1	17.0
HI (removing PSS) [42]	1690	20.3	69.6
Formamide (removing PSS) [46]	2929	17.4	88.7

4.4 The Synthesis and TE Performance Optimization of S-PEDOT Film

4.4.1 Crystalline Structure of S-PEDOT

The discovery of S-PEDOT was 1 year earlier than PEDOT:PSS. The study of S-PEDOT begin with PEDOT:FeCl$_4$, which was synthesized via the chemical oxidation of EDOT with FeCl$_3$. Unlike PEDOT:PSS, S-PEDOT is insoluble in all solvents, which can only slowly decompose in nitric acid. The performance of S-PEDOT strongly depends on its synthesis and the synthesis procedure of S-PEDOT is still a great challenge. Nevertheless, S-PEDOTs have potentially higher TE performance than that of PEDOT:PSS due to their more optimized structures.

As a polymer with good planarity and low steric effect, PEDOT can form crystalized structure when doped with small-sized anions. Gueye et al. had examined the structure of PEDOT:trifluoromethanesulfonate (PEDOT-OTf), which had high intensity diffraction peaks indicating high crystallinity of the polymer (Fig. 4.3a, b) [52]. Such crystallized region can be clearly seen in high resolution transmission electron microscope (HRTEM) images (Fig. 4.3c, d). The σ of the material is as high as 5400 S/cm [53].

The high structural ordering is beneficial to the transport properties of the material and leads to significant enhancement on TE properties [27]. In highly disordered conducting polymer such as PEDOT:PSS, the carrier near the Fermi

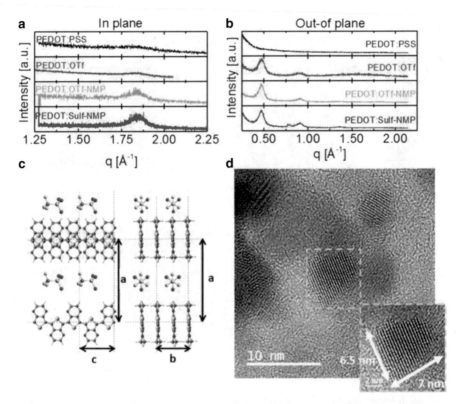

Fig. 4.3 (**a**) In-plane and (**b**) out-of plane GIWAXS diffractograms of PEDOT films. (**c**) Crystalline structure and (**d**) HRTEM image of PEDOT:OTf film [52]. Reproduced with permission

level is localized. Therefore, a highly positive temperature dependence of σ can be observed as a result of the thermal excitation. While in a much more ordered conducting polymer such as PEDOT:Tos, the carriers are delocalized and the mobility increased (Fig. 4.4a). The proportion of non-VRH conduction increases and sometimes negative temperature coefficient of σ can be achieved. The σ and S can be enhanced simultaneously when the structural order is improved (Fig. 4.4b, c). For example, if the synthesis PEDOT:Tos is stabilized with inhibitors such as poly(ethylene glycol)-block-poly(propylene glycol)-block-poly(ethylene glycol) (PEPG), the TE performance is greatly improved accompanying with the simultaneous increase of σ and S.

4.4.2 Developments in S-PEDOT's Synthesis

The chemical oxidation method is most widely used in synthesizing S-PEDOT films. The most pristine way to synthesize S-PEDOT can be found in early patents,

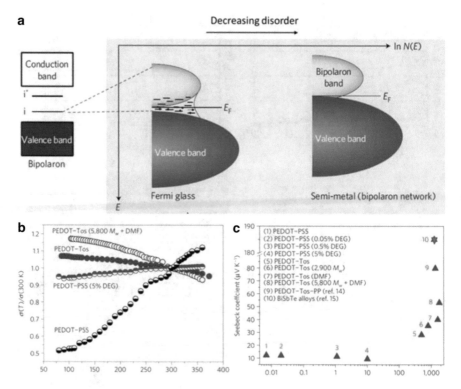

Fig. 4.4 (a) The influence of structural disorder on the band structures. (b) Temperature dependence of the electrical conductivity of PEDOT. (c) Comparison of TE performance between PEDOT films [27]. Reproduced with permission

where iron salt oxidant is directly mixed with EDOT monomer and heated to react. Randomness in TE performance is observed. Some product such as PEDOT:Tos exhibits significantly lower performance compared to those synthesized using the up-to-date technique, while some other product such as PEDOT:CSA exhibits σ of $\sim 10^3$ S/cm. The main reason of such random performance is the uncontrollable reaction rate. It was observed that the color of the reaction mixture was turning dark even before the heating process, which indicates the formation of oligomers at low temperature. According to early reports on PEDOT's polymerization mechanism, the reaction rate consists of two parts [54]. The first part is the main reaction rate (r_1) influenced by the rate determining step (RDS):

$$r_1 = k_1 \left[\mathrm{Fe}^{3+} \right]^2 [\mathrm{EDOT}]^2$$

The second part is the rate acceleration by the catalyzation effect of protons:

$$r_2 = k_2 \left[\mathrm{H}^+ \right] \left[\mathrm{Fe}^{3+} \right] [\mathrm{EDOT}]$$

When the protons are continuously generated during the reaction, the r_2 increased, which results in uncontrollable reaction rate and unqualified film with defects and oligomers inside.

To solve the above problems, base inhibitors such as imidazole or pyridine are found to be effective in reducing the reaction rate. Especially, in 2005, Jansen et al. found that the addition of pyridine in the oxidant prior to the reaction can effectively reduce the reaction rate and increase pot life [55]. Compared to some inhibitors with high boiling points that sometime stop the reactions, pyridine can be easily evaporated during the reaction, which ensured a relatively smooth reaction. The resulted PEDOT:Tos via base inhibited polymerization exhibited a high σ of ~1000 S/cm. However, the use of pyridine still cannot ensure fully stabilized reaction. New problem also emerged when vapor phase polymerization (VPP) method is used in order to produce more qualified films.

Early in 2004, Kim et al. found that EDOT can be reacted with coated oxidant layer in the form of vapor [56]. Therefore, the real-time EDOT concentration in the oxidant layer as well as the reaction rate can be effectively reduced, and the pot-life is prolonged. However, the synthesized PEDOT film exhibits a poor σ of ~60 S/cm. Fabretoo et al. have studied the reaction mechanism of vapor phase polymerization of S-PEDOT [57]. They conclude that small amount of water is essential for the transport of proton during the reaction. If the oxidant is completely dehydrated, the reaction will not start at all. Actually, in most synthesizing conditions, the oxidant and solvent will absorb certain amount of water from the environment and there will not be water shortage. However, the existence of water leads to the precipitation of oxidant in the form of hydrate during the reaction when the solvent evaporates, especially in vapor phase polymerization where solvent often evaporates prior to the completion of the reaction. When the oxidant crystallized, the crystallized part will not react with EDOT monomer and coexisted with the as-formed PEDOT (Fig. 4.5). After the film was washed, the unreacted part will form pin-hole defects in the film, which results in the deterioration of electrical properties.

The solution of crystallization problem is to add another copolymer inhibitor to the oxidant. These inhibitors have solvent preserving and anti-crystallization

Fig. 4.5 Incomplete reaction caused by the hydration (crystallization) of oxidant [57]. Reproduced with permission

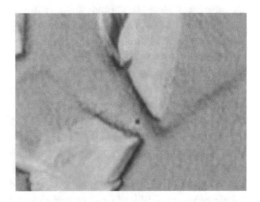

function. For example, when poly(ethylene glycol-ran-propylene glycol) (PEG-ran-PPG) is added to the oxidant, the σ of the resulted film is increased by an order of magnitude, reaching 700 S/cm. When PEPG is used as inhibitor in vacuum vapor phase polymerization (VVPP), the σ of the film may reach 1500–3500 S/cm [58–59]. Wang et al. had achieved similar σ using PEPG-inhibited VPP at moderate pressure [60]. Furthermore, the use of PEG in PEDOT can improve its biocompatibility [61–62] and electrode durability [63].

Copolymers such as PEPG not only have anti-crystallization function, but can also reduce the reaction rate due to its steric or dilution effect. Park et al. had succeeded in synthesizing high performance PEDOT film via a solution casting polymerization inhibited with PEPG/pyridine. The reported pristine PEDOT film exhibits high σ of 1354 S/cm [64]. After these successes, the combined use of base/copolymer inhibitors becomes a popular way to improve the electrical performance of S-PEDOT. However, the copolymer inhibitors will partly remain in PEDOT, which is confirmed by either differential scanning calorimeter (DSC) [63] or X-ray photoelectron spectroscopy (XPS) [60]. And the film thickness of the resultant film is decreased by ~50% due to the concentrating effect of copolymer [65]. The σ often decreases with increasing film thickness due to unequal evaporation of pyridine [64, 66–67]. As a result, such S-PEDOT film polymerized using inhibitors with high performance often has low film thickness <150 nm.

The performance enhancement can also be realized through self-inhibited polymerization (SIP) using weakly basic anions with certain steric effect, where the reaction is stabilized without the help of inhibitors. The PEDOT: camphorsulfonate (PEDOT:CSA) in early patent is a good example [68]. Shi et al. systematically studied the role of anion in SIP [69]. The anions with mediate basicity and high solubility such as CSA and dodecylbenzenesulfonate (DBSA) may serve both as base inhibitor and anti-crystallization inhibitor. High σ of ~1100 S/cm is achieved. Furthermore, the oxidant concentration may exceed 78 wt% and such method can prevent the negative effect of inhibitors on the film thickness. The film thickness is increased by nearly an order of magnitude (~2 μm) compared to the film polymerized via conventional method. However, the selection of anion is quite strict. The basicity (may partially destabilize the bipolaron) and steric effect (may affect the structural order) of anions probably impede PEDOT's electrical properties.

There are also other methods synthesizing S-PEDOT such as solution polymerization or electrochemical polymerization.

Solution polymerization often results in powdered PEDOT with poor performance. However, recent progress has been achieved using this convenient route. For example, Zhang et al. have synthesized superfine PEDOT nanowire using solution oxidation polymerization stabilized by SDS [70]. The film acquired by direct filtering exhibits high PF of 35.8 $\mu W/mK^2$.

Electrochemical polymerization is an old way of polymerizing heterocyclic conducting polymer such as polythiophene and polypyrrole. Theoretically, all conducting polymers that need to be polymerized through oxidation reaction can use this route. Electrochemically polymerized S-PEDOT has relatively severer requirement on skill and devices and the performance of the resultant films is not competitive compared to those polymerized by chemical oxidation. However, the electrochemical polymerization can be used to synthesize PEDOT with unique morphologies for some potential applications. For example, Taggart et al. polymerized PEDOT nanowires on lithographically patterned Ni nanowires via electrochemical deposition [71]. The PF $\sim$ 12 μW/mK^2 of the nanowire is measured using pre-deposited micro-electrodes. 3D microporous PEDOT material can be synthesized using a similar method. Furthermore, some 3D microporous structure [72] or network [73] which are hard to be synthesized by conventional method can also be synthesized via electrochemical polymerization.

The conventional electrochemical polymerization uses some small-sized anion such as ClO$_4^-$ as counter-ion and the TE performance of the material is relatively low compared to chemically synthesized PEDOTs. Cluebras et al. had studied the influence of counter-ion on the electropolymerized S-PEDOTs [74]. The relatively large anion can prevent the polymer chain from being coiled. The TFSI doped PEDOT using 1-Butyl-3-methylimidazolium bistrifluoromethanesulfonimide (BMIM$^+$TFSI$^-$) solution as electrolyte exhibited the highest PF of 41.6 μW/mK^2, which was further enhanced up to 147.2 μW/mK^2 via dedoping treatment. Similarly, using chemical oxidation method, Zhang et al. synthesized TFSI doped P3HT using FeTFSI$_3$ as oxidant and a high PF of $\sim$20 μW/mK^2 is achieved [2].

4.5 Dedoping Treatments

The TE performance of PEDOT can be further optimized through dedoping treatments. Basically there are two dedoping method: chemical method (including vapor and solution treatment) and electrochemical method. Different with the removal of excess PSS in the optimization of PEDOT:PSS, the dedoping aims to the removal of effectively doped anions, which results in optimized carrier concentration and PF.

Crispin et al. first conducted reduction on PEDOT:Tos [75]. In their report, PEDOT:Tos films are exposed under tetrakis(dimethylamino)ethylene (TDAE) vapor for different amount of time to achieve different doping level. When the doping level is decreased to 22%, an optimized PF and ZT value of $\sim$324 μW/mK^2 and 0.25, respectively, are obtained (Fig. 4.6a). Two years later, Park et al. reported a electrochemical dedoping treatment on PEDOT:Tos and an extremely high PF of 1270 μW/mK^2 is achieved [64]. After that, Wang reported a solution treatment method. When PEDOT:Tos is dedoped with NaBH$_4$ [60], the PF can be optimized up to 98.1 μW/mK^2 using this relatively facile method (Fig. 4.6b). Shi et al. used acidified phenylhydrazine (PHZ) to dedope PEDOT:DBSA [69], achieving a PF of 77.6 μW/mK^2. The reducing ability of the solution can be tuned via changing the acid/PHZ ratio. The change of doping level often results in the change in carrier types: bipolaron $\rightarrow$ polaron $\rightarrow$ neutral. The color of the film usually turns dark as the absorbance band moves from near-infrared area to visible light area, as shown in Fig. 4.6c, d.

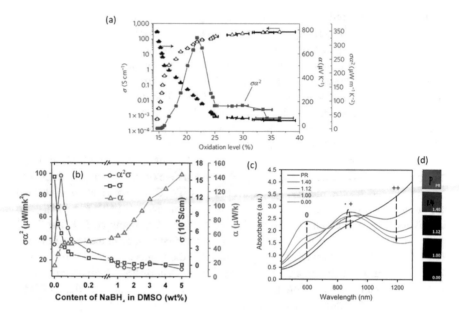

Fig. 4.6 (**a, b**) TE properties of the PEDOT–Tos films treated with (**a**) TDAE vapor [75] and (**b**) NaBH4 solution [60]. (**c**) UV–Vis–NIR spectra of pristine PEDOT:DBSA and reduced PEDOT with acid/PHZ ratio varied from 1.4 to 0. "0," "+," and "++" stand for neutral state, polaron state, and bipolaron state, respectively. (**d**) Changing in color of the reduced film [69]. Reproduced with permission

Table 4.3 The reported TE properties of PEDOT (performance optimization through controlling the carrier concentration)

Material	Dedoping method	σ (S/cm)	S (μV/K)	PF (μW/mK^2)
PEDOT:Tos [75]	TDAE vapor	~70	~210	308
PEDOT:Tos-PEPG [64]	Electrochemical	923	117	1270
PEDOT:PSS [37]	Hydrazine+UF	677	41	115
PEDOT:PSS [36]	Hydrazine solution	1310	49.3	318
PEDOT:TFSI [74]	Hydrazine solution	1100	37	147
PEDOT:Tos-PEPG [60]	NaBH$_4$ solution	~450[a]	~62[a]	98.1(165[a])
PEDOT:DBSA [69]	Acidified PHZ solution	450	35	77.2
PEDOT:FeCl$_4$ [45]	Na$_2$SO$_3$ solution	16.9	23.2	0.91
PEDOT:PSS [76]	NaOH solution	820	15.5	19.6
PEDOT:PSS [78]	NaOH solution	598.2	23.5	33.04
PEDOT:Tos-PEPG [77]	Acidity control	~460	~23	23

[a]Measured under 385 K

Similar methods are also proved effective of PEDOT:PSS. Besides, the effective doping level can also be tuned via alkali treatments [76–78], where the doping anion is substituted with immovable hydroxyl group (–OH). Table 4.3 shows a summary of the reports regarding dedoping treatments.

4.6   PEDOT-Based Nanocomposite Film

PEDOT-based nanocomposite has always been a hotspot in PEDOT-based research. The number of the reports on PEDOT-based nanocomposite TE material is almost 20 times higher than that on pure PEDOT. The selection of nanocomposites covers a wide area including carbon materials (single/multi-wall carbon nanotubes [79–86], graphene [87–97], carbon black [98–100], expanded graphite [101], mixed carbon materials [102–103]), conductive metals (Au [104–106], Ag [107–110]), Bi_2Te_3 series [68, 111–114], (PbTe [115], Bi_2S_3 [116], Sb_2Te_3 [117], Ag_2Te [118], Te-$Cu_{1.75}$Te heterostructure [119], Te-Bi_2Te_3 heterostructure [120]), oxides TE materials ($Ca_3Co_4O_9$ [121], $(Ca_{0.85}Ag_{0.15})_3Co_4O_9$ [122], Fe_2O_3 [123]), other types of semiconductive materials (Si [124], Ge [125], Te [126–135], MoS_2 [136], TiO_2 [137], ZnO [138], V_2O_5 [139–140], GeO_2 [141], BN [142], WS_2 [143], $MoSe_2$ [144], SnS [145], Cu_2SnSe_3 [146]), organic materials including other conducting polymers (polyaniline [147–150], ionic liquids [151–153], ammonium formate [154], urea [155], paper [156–157], polyester [158], and TEMPO-OH [159]), etc.

4.6.1   In-Situ Synthesis

In most of these reports, PEDOT:PSS is used as polymer matrix owing to its good water processability. Direct mixing of PEDOT and composite material results in low uniformity, oxidation of nanoparticles, and unsatisfied performance. In-situ reaction can achieve relatively more homogeneous film compared to direct mixing. Urban's groups have synthesized various PEDOT-based nanocomposite film containing Te nanowires [126–129] or Te-based heterostructure [119]. Reduction of Na_2TeO_3 using ascorbic acid in PEDOT:PSS solution results in well-dispersed Te nanowire (NW) in the polymer matrix (Fig. 4.7a). The in-situ synthesis can also prevent Te NW from oxidation. The ZT at RT reaches ~ 0.1. When $CuCl_2$ is used in combination with Na_2TeO_3, Te-$Cu_{1.75}$Te heterostructure can be formed (Fig. 4.7b) [119]. Similarly, Bae et al. synthesized Te-Bi_2Te_3 heterostructure using $BiCl_3$ instead of copper salt (Fig. 4.7c) [120]. This method provides new strategies for the engineering of carrier filtering ability of nanocomposite materials.

4.6.2   Particle Size Control

High inorganic loading often causes deterioration of flexibility as well as increasing cost. Achieving higher TE performance at relatively low inorganic lowing is still a challenge. Reports show that the effectiveness of inorganic loading can be improved through particles sized control. For example, Du et al. synthesized 3–4 nm wide p-type Bi_2Te_3 nanosheets and mixed them with PEDOT:PSS solutions [112]. The

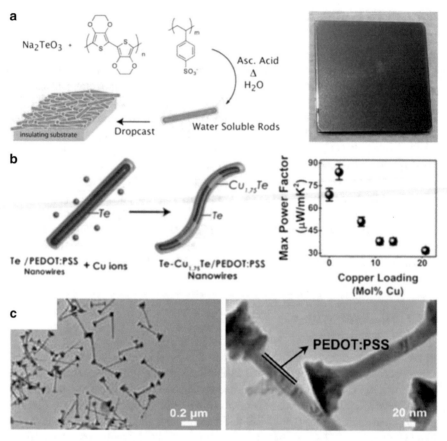

Fig. 4.7 (a) Illustration of in-situ synthesis of PEDOT:PSS-Te nanorod composite film [128]. (b) Illustration and the power factors of in-situ synthesis of PEDOT:PSS-$Cu_{1.75}$Te nanorod composite films [119]. (c) Te-$Cu_{1.75}$Te heterostructures embedded in PEDOT:PSS [120]. Reproduced with permission

resultant film exhibits a quintupled PF of 32 μW/mK^2 at a low Bi_2Te_3 loading of 4.1 wt%. Small-sized nanoparticles can create nano-sized interfaces within the polymer matrix, which can effectively scatter the low energy carriers. Such energy-filtering effect can lower the carrier concentration and improve the carrier mobility. Recently, through modifying the particles sizes/morphologies, great progresses are achieved in producing high performance PEDOT-based nanocomposites with low inorganic loadings. For example, the PEDOT films containing GeO_2 nanoparticles [141], MoS_2 nanosheets [136], and BN nanosheets [142] all exhibit PF > 45 μW/mK^2 at a low inorganic loading of <5 wt%.

4.6.3 One-Step Synthesis Using Multi-Functional Oxidants

Synthesis of PEDOT requires oxidation environment while synthesis of inorganic nanoparticles usually requires a reducing environment. That means that in most circumstances, PEDOT and inorganic nanoparticles have to be synthesized separately. Which often cause the oxidation of inorganic particles [25] in direct mixing or the reduction of PEDOT in in-situ polymerization [128]. Using multi-functional oxidant can solve this problem and realize one-step synthesis of PEDOT-based nanocomposites. For example, Zhou et al. use graphene oxide as oxidant, and synthesized PEDOT-graphene hydrogel for supercapacitors [160]. However, the σ is only 0.73 S/cm. Xu et al. used similar method and the PEDOT-graphene composite exhibits a PF of 5.2 μW/mK^2 [161]. Wang et al. use AgNO$_3$ and Cu(NO$_3$)$_2$ to synthesize PEDOT-Ag and PEDOT-Cu nanocomposites [107]. The maximum PF is up to 12.47 μW/mK^2. Similar composite materials can also be synthesized via electrochemical polymerization [162] or CVD [163]. Recently, Shi et al. synthesized PEDOT-Te nanocomposite film using TeCl$_4$ in combination with a SIP oxidant FeDBSA$_3$ [164]. The film exhibits PFs of ~100 μW/mK^2 at Te content around ~2.1–5.8 wt% (Fig. 4.8). However, the strict selection of oxidant to meet the

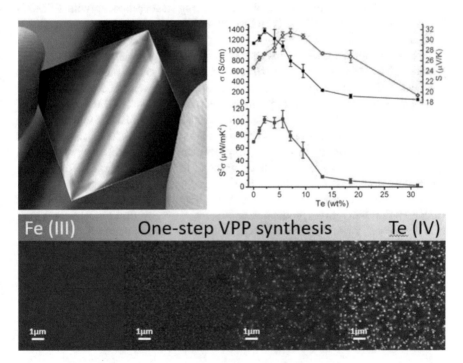

Fig. 4.8 The film morphology and TE performance of PEDOT-Te nanocomposite film polymerized under various oxidant ratio s(TeCl$_4$/FeDBSA$_3$) [164]. Reproduced with permission

requirements of good solubility as well as enough oxidizing ability has limited the application of multi-functional oxidant towards more material systems (e.g., Ag-low solubility, Bi and Sb-low oxidizing potentials, etc.). Table 4.4 summarizes the TE properties of PEDOT-based nanocomposite films.

4.7   Future Prospective on PEDOT-Based TE Films

According to the calculation, PEDOT matrix has theoretically ultra-high σ and S [165]. That means there is still much space for the improvement of its TE performance. The development of new synthesizing skill or post-treatment may lead to more ordered polymer structures and further enhanced TE properties. In addition, through molecular design, novel monomers may also enhance the chain conductance of the polymer.

Furthermore, the PEDOT-based nanocomposites have even larger potentials for performance optimization. The discovery of new composite, new multi-functional oxidant or new synthesizing techniques may leads to improvement on particle distribution, particle size/morphology control, and protection against oxidation. New material system that is suitable for building nanocomposite with PEDOTs needs to be explored. Especially, p-type materials with high S and reasonable σ are desired.

Various mechanisms still remain poorly understood, especially on the synthesis, doping behavior as well as the performance enhancement of nanocomposite. For example, about anion doping, the structures of PEDOTs with >25% doping levels cannot be fitted into the currently proposed structural models. The mechanism of dedoping by alkali is also unclear. The reaction mechanism of EDOT with non-ferric oxidant has not been answered. Theories about nanocomposites such as energy filtering or phonon dragging still need improvement through either characterization or calculation. We may have a better control of PEDOT's synthesis and structure when these problems are cleared.

Furthermore, for practical TE applications, improvements on various properties (radiation durability, mass production ability, transparency, film thickness, etc.) are also demanded. More work has to be done on this promising material.

Table 4.4 TE properties of reported PEDOT-based composite materials

Composites	Loading	Stabilizer/treatment	σ (S/cm)	S (μV/K)	PF (μW/mK^2)
PEDOT:PSS/Au nanoparticle [104]	$\sim10^{-5}$ wt%	DT	241	22	11.68
PEDOT:PSS/Au nanoparticle [105]	$\sim10^{-5}$ wt%	MHA	730	26.5	51.2
PEDOT:PSS/GeO$_2$ nanoparticle [141]	0.1 wt%	PVP	$\sim1050^a$	$\sim19^a$	47^a
PEDOT:PSS/graphene [87]	2 wt%		58.77	32.13	11.09
PEDOT:Cl/Te quantum dots [164]	2.1 wt%		1383	27.4	103.5
PEDOT:PSS/carbon black [98]	2.5 wt%		61.2	~12.5	0.96
PEDOT:PSS/rGO [88]	3 wt%		637	26.8	45.68
PEDOT:PSS/rGO [89]	3 wt%		~1160	~16.8	32.6
PEDOT:PSS/rGO [93]	3 wt%	Reduction	519	30.2	47.4
PEDOT:PSS/graphene [90]	3 wt%		1283	22.0	38.6
PEDOT:PSS/MoS$_2$ NS [136]	4 wt%		1229	19.5	45.6
PEDOT:PSS/Bi$_{0.5}$Te$_{3.0}$Sb$_{1.5}$ NS [112]	4.1 wt%		~1300	~15.8	32.26
PEDOT:PSS/graphene+MWCNT [102]	5 wt%		689	23.2	37.08
PEDOT:PSS/((Ca$_x$Ag$_y$)$_3$Co$_4$O$_9$ [122]	5 wt%	AT	2455	17.5	75.2
PEDOT:PSS/BN NS + DMSO [142]	5 wt%		1979	22.5	100.1
PEDOT:PSS/2D MoSe$_2$ NS [144]	5 wt%	Butyllithium	1143	20.6	48.6
PEDOT:PSS/ SnS nanobelt [145]	8 wt%		~760	19.1	27.8
PEDOT:PSS/Ca$_3$Co$_4$O$_9$ [121]	9.1 wt%		–	–	~3.8
PEDOT:PSS/ammonium formate [154]	10 wt%		0.036	436.3	0.49
PEDOT:PSS/Bi$_2$Te$_3$ particle [111]	10 wt%		421	~15	9.9
PEDOT:Tos/carbon black [100]	10 wt%	DBSA	~43	~15	0.993
PEDOT:PSS/Bi$_2$Te$_3$ NW [114]	10 wt%		401	16.3	10.6
PEDOT:PSS/rGO + SWCNT [103]	>10 wt%	AT	208.4	20.9	9.1
PEDOT:PSS/rGO [161]	16 wt%		50.8	31.8	5.2

PEDOT:PSS/graphene [97]	17 wt%		~150	~30	14.2
PEDOT:PSS/Ag NW [108]	20 wt%	PVP	84.84	18.3	2.86
PEDOT:Tos/MWCNT [82]	26.5 wt%		~83	56.1	25.9
PEDOT:PSS/Ge powder [125]	29.6 wt%		~600	~50	165
PEODT:FeCl$_4$/PbTe nanoparticle [115]	30 wt%		6.4×10^{-4}	−4088	1.4
PEDOT:PSS/carbon black [99]	30 wt%		20.34	7.1	0.10
PEDOT:PSS/SWCNT [79]	35 wt%	Gum Arabic	~400	~25	~25
PEDOT:Tos-PEPG/SWCNT [96]	35 wt%	Reduction	318	37.2	44.1
PEDOT:PSS/Bi$_2$S$_3$ nanotube [116]	36.1 wt%		0.013	−1318	2.3
PEDOT:PSS/WS$_2$ NS + DMSO [143]	50 wt%		1025	21	45.2
PEDOT:PSS/Bi$_2$Te$_3$ + graphene [92]	~50 wt%		400	20	160
PEDOT:PSS/ Te-Cu$_{1.75}$Te heterostructure [119]	~50 wt%		–	–	84
PEDOT:PSS/ (EMIM)BF$_4$ [152]	50 vol%		~750	~23	38.46
PEDOT:PSS/SWCNT [80]	60 wt%	PVAc	~1000	~40	~160
PEDOT:Tos-PEPG/SWCNT [81]	66.7 wt%		78.6	46.3	16.8
PEDOT:PSS/(BMIM)Br [151]	70 wt%		~125	~28	9.9
PEDOT:PSS/graphene [137]	70 wt%		~1800	~10	~17.5
PEDOT:PSS/MWCNT [85]	~70 wt%		~7.5	~14	0.229
PEDOT:PSS/PEDOT-Te nanorod [133]	~70 wt%		~680	27.5	51.4
PEDOT:PSS/SWCNT [83]	74 wt%		3800	28	300
PEDOT:Bzs/graphene [91]	75 wt%	DBSA	~150	~43	~28
PEDOT:PSS/Au nanorod [106]	75 wt%	PEG-thiol	~2000	~12	28.8
PEDOT:PSS/expanded graphite [101]	80 wt%		213	~16	5.31
PEDOT:PSS/Te nanorod [131]	80 wt%	AT	~500	~29	42.1

(continued)

Table 4.4 (continued)

Composites	Loading	Stabilizer/treatment	σ (S/cm)	S (μV/K)	PF (μW/mK^2)
PEDOT:PSS/Te + SWCNT [134]	~80 wt%		139	118	206
PEDOT:PSS/Te NW [127]	~84 wt%		–	–	~100
PEDOT:PSS/Te NW [128]	85 wt%		19.3	163	70.9
PEDOT:PSS/Te NW [126]	~85 wt%		11	180	~35
PEDOT:PSS/p-type Bi$_2$Te$_3$ [25]	90 vol%		~145	~70	131
PEDOT:PSS/Bi$_{0.4}$Te$_{3.0}$Sb$_{1.6}$ r [68]	95 wt%	PAA	380	79	237
PEDOT:PSS/GINC [123]	95 wt%		~800	~25.5	51.93
PEDOT:PSS/V$_2$O$_5$ [139]	>95 wt%		~0.05	~ – 350	0.16
PEDOT:PSS/ Te-Bi$_2$Te$_3$ heterostructures [120]	>>50 wt%		8.21	141.02	16.33
PEDOT:PSS/Te nanorod [132]	>>50 wt%	AT	214.86	114.97	284
PEDOT:PSS/paper +30 wt% EG [156]	>>50 wt%		54.1	~22.5	3.0
PEDOT:PSS/Sb$_2$Te$_3$ [117]	>>50 wt%		~126	132	~220
PEDOT:ClO$_4$/MWCNT electropolymerized [94]	In-situ deposition	DOC etc.	2100	~27	155
PEDOT:PSS /PANI:CSA [147]	Layers		1550	19	56
PEDOT:PSS/RTCVD graphene [95]	Covered		1096	28.1	57.9
PEDOT:PSS/ZnO flower [138]	Covered		~9	~21	0.4
PEDOT:PSS/Si [124]	2 layers		~49	73	26.2

DT dodecanethiol, *NS* nanosheets, *rGO* reduced graphene oxide, *MHA* 6-mercaptohexanoic acid, *PVAc* polyvinyl acetate, *PAA* polyacrylic acid, *SWCNT* single-wall carbon nanotube, *PEG-thiol* PEG modified with thiol groups, *EMIM* 1-ethyl-3-methylimidazolium, *GTNC* graphene–TiO$_2$ nanocomposite, *Bzs* benzenesulfonate, *GINC* graphene–iron oxide nanocomposite, *DOC* sodium deoxycholate, *RTCVD* rapid thermal chemical vapor deposition, *AT* acid treatment
Notes: 1. When codoping exists, only the most competitive anion is listed. 2. The use of solvent treatment of DMSO, DMF, and NMP is not listed as additives
^{a}Measured under 388 K

References

1. G. Tourillon, F. Garnier, New electrochemically generated organic conducting polymers. J. Electroanal. Chem. Interfacial Electrochem. **135**(1), 173–178 (1982)
2. Q. Zhang, Y. Sun, W. Xu, D. Zhu, Thermoelectric energy from flexible P3HT films doped with a ferric salt of triflimide anions. Energy Environ. Sci. **5**(11), 9639 (2012)
3. C. Bounioux, P. Díaz-Chao, M. Campoy-Quiles, M.S. Martín-González, A.R. Goñi, R. Yerushalmi-Rozen, C. Müller, Thermoelectric composites of poly(3-hexylthiophene) and carbon nanotubes with a large power factor. Energy Environ. Sci. **6**(3), 918 (2013)
4. Y. Du, K.F. Cai, S.Z. Shen, P.S. Casey, Preparation and characterization of graphene nanosheets/poly(3-hexylthiophene) thermoelectric composite materials. Synth. Met. **162**(23), 2102–2106 (2012)
5. S. Qu, Q. Yao, L. Wang, Z. Chen, K. Xu, H. Zeng, W. Shi, T. Zhang, C. Uher, L. Chen, Highly anisotropic P3HT films with enhanced thermoelectric performance via organic small molecule epitaxy. NPG Asia Mater. **8**(7), e292–e292 (2016)
6. M. He, J. Ge, Z. Lin, X. Feng, X. Wang, H. Lu, Y. Yang, F. Qiu, Thermopower enhancement in conducting polymer nanocomposites via carrier energy scattering at the organic–inorganic semiconductor interface. Energy Environ. Sci. **5**(8), 8351 (2012)
7. F. Jonas, G. Heywang, W. Schmidtberg, DE 3813589A1, 1988
8. F. Jonas, G. Heywang, W. Schmidtberg, DE 3814730A1, 1988
9. C. Zhang, T.M. Higgins, S.-H. Park, S.E. O'Brien, D. Long, J.N. Coleman, V. Nicolosi, Highly flexible and transparent solid-state supercapacitors based on RuO2/PEDOT:PSS conductive ultrathin films. Nano Energy **28**, 495–505 (2016)
10. Y. Yang, L. Zhang, S. Li, W. Yang, J. Xu, Y. Jiang, J. Wen, Electrochemical performance of conducting polymer and its nanocomposites prepared by chemical vapor phase polymerization method. J. Mater. Sci. Mater. Electron. **24**(7), 2245–2253 (2013)
11. T. Cheng, Y.-Z. Zhang, J.-D. Zhang, W.-Y. Lai, W. Huang, High-performance free-standing PEDOT:PSS electrodes for flexible and transparent all-solid-state supercapacitors. J. Mater. Chem. A **4**(27), 10493–10499 (2016)
12. D. Aradilla, F. Gao, G. Lewes-Malandrakis, W. Muller-Sebert, P. Gentile, M. Boniface, D. Aldakov, B. Iliev, T.J. Schubert, C.E. Nebel, G. Bidan, Designing 3D multihierarchical heteronanostructures for high-performance on-chip hybrid supercapacitors: poly(3,4-(ethylenedioxy)thiophene)-coated diamond/silicon nanowire electrodes in an aprotic ionic liquid. ACS Appl. Mater. Interfaces **8**(28), 18069–18077 (2016)
13. W. Zhang, B. Zhao, Z. He, X. Zhao, H. Wang, S. Yang, H. Wu, Y. Cao, High-efficiency ITO-free polymer solar cells using highly conductive PEDOT:PSS/surfactant bilayer transparent anodes. Energy Environ. Sci. **6**(6), 1956 (2013)
14. H. Tian, Z. Yu, A. Hagfeldt, L. Kloo, L. Sun, Organic redox couples and organic counter electrode for efficient organic dye-sensitized solar cells. J. Am. Chem. Soc. **133**(24), 9413–9422 (2011)
15. M. Zheng, J. Huo, B. Chen, Y. Tu, J. Wu, L. Hu, S. Dai, Pt–Co and Pt–Ni hollow nanospheres supported with PEDOT:PSS used as high performance counter electrodes in dye-sensitized solar cells. Sol. Energy **122**, 727–736 (2015)
16. S. Radhakrishnan, C. Sumathi, A. Umar, S. Jae Kim, J. Wilson, V. Dharuman, Polypyrrole-poly(3,4-ethylenedioxythiophene)-Ag (PPy-PEDOT-Ag) nanocomposite films for label-free electrochemical DNA sensing. Biosens. Bioelectron. **47**, 133–140 (2013)
17. J.-C. Wang, R.S. Karmakar, Y.-J. Lu, M.-C. Wu, K.-C. Wei, Nitrogen plasma surface modification of poly(3,4-ethylenedioxythiophene):poly(styrenesulfonate) films to enhance the piezoresistive pressure-sensing properties. J. Phys. Chem. C **120**(45), 25977–25984 (2016)
18. B. Le Ouay, M. Boudot, T. Kitao, T. Yanagida, S. Kitagawa, T. Uemura, Nanostructuration of PEDOT in porous coordination polymers for Tunable porosity and conductivity. J. Am. Chem. Soc. **138**(32), 10088–10091 (2016)

19. T. Vuorinen, J. Niittynen, T. Kankkunen, T.M. Kraft, M. Mantysalo, Inkjet-printed Graphene/PEDOT:PSS temperature sensors on a skin-conformable polyurethane substrate. Sci. Rep. **6**, 35289 (2016)
20. T. Park, J. Na, B. Kim, Y. Kim, H. Shin, E. Kim, Photothermally activated pyroelectric polymer films for harvesting of solar heat with a hybrid energy cell structure. ACS Nano **9**(12), 11830–11839 (2015)
21. F. Jonas, W. Krafft, EP 440957, 1990
22. F. Jiang, J. Xu, B. Lu, X. Yu, R. Huang, L. Li, Thermoelectric performance of poly(3,4-ethylenedioxythiophene):poly(styrenesulfonate). Chin. Phys. Lett. **25**(6), 2202 (2008)
23. K.-C. Chang, M.-S. Jeng, C.-C. Yang, Y.-W. Chou, S.-K. Wu, M.A. Thomas, Y.-C. Peng, The thermoelectric performance of poly(3,4-ethylenedi oxythiophene)/poly(4-styrenefonate) thin films. J. Electron. Mater. **38**(7), 1182–1188 (2009)
24. Z. Fan, D. Du, H. Yao, J. Ouyang, Higher PEDOT molecular weight giving rise to higher thermoelectric property of PEDOT:PSS: a comparative study of clevios P and clevios PH1000. ACS Appl. Mater. Interfaces **9**(13), 11732–11738 (2017)
25. B. Zhang, J. Sun, H.E. Katz, F. Fang, R.L. Opila, Promising thermoelectric properties of commercial PEDOT:PSS materials and their bi2Te3 powder composites. ACS Appl. Mater. Interfaces **2**(11), 3170–3178 (2010)
26. M. Scholdt, H. Do, J. Lang, A. Gall, A. Colsmann, U. Lemmer, J.D. Koenig, M. Winkler, H. Boettner, Organic semiconductors for thermoelectric applications. J. Electron. Mater. **39**(9), 1589–1592 (2010)
27. O. Bubnova, Z.U. Khan, H. Wang, S. Braun, D.R. Evans, M. Fabretto, P. Hojati-Talemi, D. Dagnelund, J.B. Arlin, Y.H. Geerts, S. Desbief, D.W. Breiby, J.W. Andreasen, R. Lazzaroni, W.M. Chen, I. Zozoulenko, M. Fahlman, P.J. Murphy, M. Berggren, X. Crispin, Semi-metallic polymers. Nat. Mater. **13**(2), 190–194 (2014)
28. U. Lang, M. Elisabeth, N. Nicola, D. Jurg, Microscopical investigations of PEDOT:PSS thin films. Adv. Funct. Mater. **19**(8), 1215–1220 (2010)
29. A.M. Nardes, M. Kemerink, R.A.J. Janssen, Anisotropic hopping conduction in spin-coated PEDOT:PSS thin films. Phys. Rev. B **76**(8), 085208 (2007)
30. S. Liu, H. Deng, Y. Zhao, S. Ren, Q. Fu, The optimization of thermoelectric properties in a PEDOT:PSS thin film through post-treatment. RSC Adv. **5**(3), 1910–1917 (2015)
31. J. Xiong, F. Jiang, W. Zhou, C. Liu, J. Xu, Highly electrical and thermoelectric properties of a PEDOT:PSS thin-film via direct dilution–filtration. RSC Adv. **5**(75), 60708–60712 (2015)
32. P. Zhao, Q. Tang, X. Zhao, Y. Tong, Y. Liu, Highly stable and flexible transparent conductive polymer electrode patterns for large-scale organic transistors. J. Colloid Interface Sci. **520**, 58–63 (2018)
33. C. Liu, B. Lu, J. Yan, J. Xu, R. Yue, Z. Zhu, S. Zhou, X. Hu, Z. Zhang, P. Chen, Highly conducting free-standing poly(3,4-ethylenedioxythiophene)/poly(styrenesulfonate) films with improved thermoelectric performances. Synth. Met. **160**(23–24), 2481–2485 (2010)
34. E. Liu, C. Liu, Z. Zhu, H. Shi, Q. Jiang, F. Jiang, J. Xu, J. Xiong, Y. Hu, Enhanced thermoelectric performance of PEDOT:PSS films by solvent thermal treatment. J. Polym. Res. **22**(12), 240 (2015)
35. G.H. Kim, L. Shao, K. Zhang, K.P. Pipe, Engineered doping of organic semiconductors for enhanced thermoelectric efficiency. Nat. Mater. **12**(8), 719–723 (2013)
36. S.H. Lee, H. Park, S. Kim, W. Son, I.W. Cheong, J.H. Kim, Transparent and flexible organic semiconductor nanofilms with enhanced thermoelectric efficiency. J. Mater. Chem. A **2**(20), 7288 (2014)
37. S.H. Lee, H. Park, W. Son, H.H. Choi, J.H. Kim, Novel solution-processable, dedoped semiconductors for application in thermoelectric devices. J. Mater. Chem. A **2**(33), 13380 (2014)
38. N. Massonnet, A. Carella, O. Jaudouin, P. Rannou, G. Laval, C. Celle, J.-P. Simonato, Improvement of the Seebeck coefficient of PEDOT:PSS by chemical reduction combined with a novel method for its transfer using free-standing thin films. J. Mater. Chem. C **2**(7), 1278–1283 (2014)

39. J. Kim, J.G. Jang, J.-I. Hong, S.H. Kim, J. Kwak, Sulfuric acid vapor treatment for enhancing the thermoelectric properties of PEDOT:PSS thin-films. J. Mater. Sci. Mater. Electron. **27**(6), 6122–6127 (2016)

40. C. Liu, H. Shi, J. Xu, Q. Jiang, H. Song, Z. Zhu, Improved thermoelectric properties of PEDOT:PSS nanofilms treated with oxalic acid. J. Electron. Mater. **44**(6), 1791–1795 (2014)

41. D.A. Mengistie, C.H. Chen, K.M. Boopathi, F.W. Pranoto, L.J. Li, C.W. Chu, Enhanced thermoelectric performance of PEDOT:PSS flexible bulky papers by treatment with secondary dopants. ACS Appl. Mater. Interfaces **7**(1), 94–100 (2015)

42. J. Wang, K. Cai, H. Song, S. Shen, Simultaneously enhanced electrical conductivity and Seebeck coefficient in poly (3,4-ethylenedioxythiophene) films treated with hydroiodic acid. Synth. Met. **220**, 585–590 (2016)

43. J. Wang, K. Cai, S. Shen, Enhanced thermoelectric properties of poly(3,4-ethylenedioxythiophene) thin films treated with H 2 SO 4. Org. Electron. **15**(11), 3087–3095 (2014)

44. C. Yi, L. Zhang, R. Hu, S.S.C. Chuang, J. Zheng, X. Gong, Highly electrically conductive polyethylenedioxythiophene thin films for thermoelectric applications. J. Mater. Chem. A **4**(33), 12730–12738 (2016)

45. J. Zhao, D. Tan, G. Chen, A strategy to improve the thermoelectric performance of conducting polymer nanostructures. J. Mater. Chem. C **5**(1), 47–53 (2017)

46. A.K.K. Kyaw, T.A. Yemata, X. Wang, S.L. Lim, W.S. Chin, K. Hippalgaonkar, J. Xu, Enhanced thermoelectric performance of PEDOT:PSS films by sequential post-treatment with formamide. Macromol. Mater. Eng. **303**(2), 1700429 (2018)

47. Z. Fan, D. Du, Z. Yu, P. Li, Y. Xia, J. Ouyang, Significant enhancement in the thermoelectric properties of PEDOT:PSS films through a treatment with organic solutions of inorganic salts. ACS Appl. Mater. Interfaces **8**(35), 23204–23211 (2016)

48. E. Yang, J. Kim, B.J. Jung, J. Kwak, Enhanced thermoelectric properties of sorbitol-mixed PEDOT:PSS thin films by chemical reduction. J. Mater. Sci. Mater. Electron. **26**(5), 2838–2843 (2015)

49. J. Kim, R. Patel, B.J. Jung, J. Kwak, Simultaneous improvement of performance and stability in PEDOT:PSS–sorbitol composite based flexible thermoelectric modules by novel design and fabrication process. Macromol. Res. **26**(1), 61–65 (2017)

50. Z. Zhu, C. Liu, H. Shi, Q. Jiang, J. Xu, F. Jiang, J. Xiong, E. Liu, An effective approach to enhanced thermoelectric properties of PEDOT:PSS films by a DES post-treatment. J. Polym. Sci. B Polym. Phys. **53**(12), 885–892 (2015)

51. Z. Zhu, C. Liu, Q. Jiang, H. Shi, J. Xu, F. Jiang, J. Xiong, E. Liu, Green DES mixture as a surface treatment recipe for improving the thermoelectric properties of PEDOT:PSS films. Synth. Met. **209**, 313–318 (2015)

52. K.E. Aasmundtveit, E.J. Samuelsen, L.A.A. Pettersson, O. Inganäs, T. Johansson, R. Fei-denhans, Structure of thin films of poly(3,4-ethylenedioxythiophene). Synth. Met. **101**(1–3), 561–564 (1999)

53. M.N. Gueye, A. Carella, N. Massonnet, E. Yvenou, S. Brenet, J. Faure-Vincent, S. Pouget, F. Rieutord, H. Okuno, A. Benayad, R. Demadrille, J.-P. Simonato, Structure and dopant engineering in PEDOT thin films: practical tools for a dramatic conductivity enhancement. Chem. Mater. **28**(10), 3462–3468 (2016)

54. A. Elschner, S. Kirchmeyer, W. Lövenich, U. Merker, K. Reuter, *Principles and Applications of an Instrinsically Conductive Polymer* (Taylor & Francis Group, New York, 2011)

55. B. Winther-Jensen, D.W. Breiby, K. West, Base inhibited oxidative polymerization of 3,4-ethylenedioxythiophene with iron(III)tosylate. Synth. Met. **152**(1–3), 1–4 (2005)

56. J. Kim, E. Kim, Y. Won, H. Lee, K. Suh, The preparation and characteristics of conductive poly(3,4-ethylenedioxythiophene) thin film by vapor-phase polymerization. Synth. Met. **139**(2), 485–489 (2003)

57. M. Fabretto, K. Zuber, C. Hall, P. Murphy, H.J. Griesser, The role of water in the synthesis and performance of vapour phase polymerised PEDOT electrochromic devices. J. Mater. Chem. **19**(42), 7871 (2009)

58. M.V. Fabretto, D.R. Evans, M. Mueller, K. Zuber, P. Hojati-Talemi, R.D. Short, G.G. Wallace, P.J. Murphy, Polymeric material with metal-like conductivity for next generation organic electronic devices. Chem. Mater. **24**(20), 3998–4003 (2012)
59. M. Mueller, M. Fabretto, D. Evans, P. Hojati-Talemi, C. Gruber, P. Murphy, Vacuum vapour phase polymerization of high conductivity PEDOT: role of PEG-PPG-PEG, the origin of water, and choice of oxidant. Polymer **53**(11), 2146–2151 (2012)
60. J. Wang, K. Cai, S. Shen, A facile chemical reduction approach for effectively tuning thermoelectric properties of PEDOT films. Org. Electron. **17**, 151–158 (2015)
61. L.H. Jimison, A. Hama, X. Strakosas, V. Armel, D. Khodagholy, E. Ismailova, G.G. Malliaras, B. Winther-Jensen, R.M. Owens, PEDOT:TOS with PEG: a biofunctional surface with improved electronic characteristics. J. Mater. Chem. **22**(37), 19498 (2012)
62. M. Mir, R. Lugo, I.B. Tahirbegi, J. Samitier, Miniaturizable ion-selective arrays based on highly stable polymer membranes for biomedical applications. Sensors **14**(7), 11844–11854 (2014)
63. B. Winther-Jensen, K. Fraser, C. Ong, M. Forsyth, D.R. MacFarlane, Conducting polymer composite materials for hydrogen generation. Adv. Mater. **22**(15), 1727–1730 (2010)
64. T. Park, C. Park, B. Kim, H. Shin, E. Kim, Flexible PEDOT electrodes with large thermoelectric power factors to generate electricity by the touch of fingertips. Energy Environ. Sci. **6**(3), 788 (2013)
65. M. Fabretto, M. Muller, K. Zuber, P. Murphy, Influence of PEG-ran-PPG surfactant on vapour phase polymerised PEDOT thin films. Macromol. Rapid Commun. **30**, 1846–1851 (2009)
66. P.A. Levermore, L. Chen, X. Wang, R. Das, D.D.C. Bradley, Fabrication of highly conductive poly(3,4-ethylenedioxythiophene) films by vapor phase polymerization and their application in efficient organic light-emitting diodes. Adv. Mater. **19**(17), 2379–2385 (2007)
67. S. Kirchmeyer, J. Freidrich, *Transparent Polythiophene Layers of High Conductivity*. US20030161941A1, 2003
68. K. Kato, H. Hagino, K. Miyazaki, Fabrication of bismuth telluride thermoelectric films containing conductive polymers using a printing method. J. Electron. Mater. **42**(7), 1313–1318 (2013)
69. W. Shi, Q. Yao, S. Qu, H. Chen, T. Zhang, L. Chen, Micron-thick highly conductive PEDOT films synthesized via self-inhibited polymerization: Roles of anions. NPG Asia Mater. **9**(7), e405 (2017)
70. J. Zhang, K. Zhang, F. Xu, S. Wang, Y. Qiu, Thermoelectric transport in ultrathin poly(3,4-ethylenedioxythiophene) nanowire assembly. Compos. Part B **136**, 234–240 (2018)
71. D.K. Taggart, Y. Yang, S.C. Kung, T.M. McIntire, R.M. Penner, Enhanced thermoelectric metrics in ultra-long electrodeposited PEDOT nanowires. Nano Lett. **11**(1), 125–131 (2011)
72. A. García-Barberá, M. Culebras, S. Roig-Sánchez, C.M. Gómez, A. Cantarero, Three dimensional PEDOT nanowires network. Synth. Met. **220**, 208–212 (2016)
73. G. Ye, J. Xu, X. Ma, Q. Zhou, D. Li, Y. Zuo, L. Lv, W. Zhou, X. Duan, One-step electrodeposition of free-standing flexible conducting PEDOT derivative film and its electrochemical capacitive and thermoelectric performance. Electrochim. Acta **224**, 125–132 (2017)
74. M. Culebras, C.M. Gómez, A. Cantarero, Enhanced thermoelectric performance of PEDOT with different counter-ions optimized by chemical reduction. J. Mater. Chem. A **2**(26), 10109 (2014)
75. O. Bubnova, Z.U. Khan, A. Malti, S. Braun, M. Fahlman, M. Berggren, X. Crispin, Optimization of the thermoelectric figure of merit in the conducting polymer poly(3,4-ethylenedioxythiophene). Nat. Mater. **10**(6), 429–433 (2011)
76. T.-C. Tsai, H.-C. Chang, C.-H. Chen, Y.-C. Huang, W.-T. Whang, A facile dedoping approach for effectively tuning thermoelectricity and acidity of PEDOT:PSS films. Org. Electron. **15**(3), 641–645 (2014)
77. Z.U. Khan, O. Bubnova, M.J. Jafari, R. Brooke, X. Liu, R. Gabrielsson, T. Ederth, D.R. Evans, J.W. Andreasen, M. Fahlman, X. Crispin, Acido-basic control of the thermoelectric properties of poly(3,4-ethylenedioxythiophene)tosylate (PEDOT-Tos) thin films. J. Mater. Chem. C **3**(40), 10616–10623 (2015)

78. Z. Zhu, C. Liu, Q. Jiang, H. Shi, F. Jiang, J. Xu, J. Xiong, E. Liu, Optimizing the thermoelectric properties of PEDOT:PSS films by combining organic co-solvents with inorganic base. J. Mater. Sci. Mater. Electron. **26**(11), 8515–8521 (2015)

79. D. Kim, Y. Kim, K. Choi, J. Grunlan, C. Yu, Improved thermoelectric behavior of nanotube-filled polymer composites with poly(3,4-ethylenedioxythiophene) poly(styrenesulfonate). ACS Nano **4**(1), 513 (2010)

80. C. Yu, K. Choi, L. Yin, J. Grunlan, Light-weight flexible carbon nanotube based organic composites with large thermoelectric power factors. ACS Nano **5**(10), 7885–7892 (2011)

81. Q. Jiang, C. Liu, D. Zhu, H. Song, J. Xu, H. Shi, D. Mo, Z. Wang, Z. Zhu, Simultaneous enhancement of the electrical conductivity and seebeck coefficient of PEDOT-block-PEG/SWCNTs nanocomposites. J. Electron. Mater. **44**(6), 1585–1591 (2014)

82. Y.Y. Wang, K.F. Cai, S. Shen, X. Yao, In-situ fabrication and enhanced thermoelectric properties of carbon nanotubes filled poly(3,4-ethylenedioxythiophene) composites. Synth. Met. **209**, 480–483 (2015)

83. L. Zhang, Y. Harima, I. Imae, Highly improved thermoelectric performances of PEDOT:PSS/SWCNT composites by solvent treatment. Org. Electron. **51**, 304–307 (2017)

84. J.-H. Hsu, W. Choi, G. Yang, C. Yu, Origin of unusual thermoelectric transport behaviors in carbon nanotube filled polymer composites after solvent/acid treatments. Org. Electron. **45**, 182–189 (2017)

85. Z. Zhang, G. Chen, H. Wang, X. Li, Template-directed in situ polymerization preparation of nanocomposites of PEDOT:PSS-coated multi-walled carbon nanotubes with enhanced thermoelectric property. Chem. Asian J. **10**(1), 149–153 (2015)

86. X. Hu, G. Chen, X. Wang, An unusual coral-like morphology for composites of poly(3,4-ethylenedioxythiophene)/carbon nanotube and the enhanced thermoelectric performance. Compos. Sci. Technol. **144**, 43–50 (2017)

87. G.H. Kim, D.H. Hwang, S.I. Woo, Thermoelectric properties of nanocomposite thin films prepared with poly(3,4-ethylenedioxythiophene) poly(styrenesulfonate) and graphene. Phys. Chem. Chem. Phys. **14**(10), 3530–3536 (2012)

88. D. Yoo, J. Kim, J.H. Kim, Direct synthesis of highly conductive poly(3,4-ethylenedioxythiophene):poly(4-styrenesulfonate) (PEDOT:PSS)/graphene composites and their applications in energy harvesting systems. Nano Res. **7**(5), 717–730 (2014)

89. F. Li, K. Cai, S. Shen, S. Chen, Preparation and thermoelectric properties of reduced graphene oxide/PEDOT:PSS composite films. Synth. Met. **197**, 58–61 (2014)

90. J. Xiong, F. Jiang, H. Shi, J. Xu, C. Liu, W. Zhou, Q. Jiang, Z. Zhu, Y. Hu, Liquid exfoliated graphene as dopant for improving the thermoelectric power factor of conductive PEDOT:PSS nanofilm with hydrazine treatment. ACS Appl. Mater. Interfaces **7**(27), 14917–14925 (2015)

91. H. Ju, M. Kim, J. Kim, Enhanced thermoelectric performance by alcoholic solvents effects in highly conductive benzenesulfonate-doped poly(3,4-ethylenedioxythiophene)/graphene composites. J. Appl. Polym. Sci. **132**(24), 42107 (2015)

92. A.A.A. Rahman, A.A. Umar, X. Chen, M.M. Salleh, M. Oyama, Enhanced thermoelectric properties of bismuth telluride–organic hybrid films via graphene doping. Appl. Phys. A Mater. Sci. Process. **122**(2), 133 (2016)

93. R. Sarabia-Riquelme, G. Ramos-Fernández, I. Martin-Gullon, M.C. Weisenberger, Synergistic effect of graphene oxide and wet-chemical hydrazine/deionized water solution treatment on the thermoelectric properties of PEDOT:PSS sprayed films. Synth. Met. **222**, 330–337 (2016)

94. M. Culebras, C. Cho, M. Krecker, R. Smith, Y. Song, C.M. Gomez, A. Cantarero, J.C. Grunlan, High thermoelectric power factor organic thin films through combination of nanotube multilayer assembly and electrochemical polymerization. ACS Appl. Mater. Interfaces **9**(7), 6306–6313 (2017)

95. C. Park, D. Yoo, S. Im, S. Kim, W. Cho, J. Ryu, J.H. Kim, Large-scalable RTCVD Graphene/PEDOT:PSS hybrid conductive film for application in transparent and flexible thermoelectric nanogenerators. RSC Adv. **7**(41), 25237–25243 (2017)

96. J. Wang, K. Cai, J. Yin, S. Shen, Thermoelectric properties of the PEDOT/SWCNT composite films prepared by a vapor phase polymerization. Synth. Met. **224**, 27–32 (2017)
97. K. Xu, G. Chen, D. Qiu, In situ chemical oxidative polymerization preparation of poly(3,4-ethylenedioxythiophene)/graphene nanocomposites with enhanced thermoelectric performance. Chem. Asian J. **10**(5), 1225–1231 (2015)
98. Y. Du, K.F. Cai, S.Z. Shen, W.D. Yang, P.S. Casey, The thermoelectric performance of carbon black/poly(3,4-ethylenedioxythiophene):poly(4-styrenesulfonate) composite films. J. Mater. Sci. Mater. Electron. **24**(5), 1702–1706 (2012)
99. L. Wang, F. Jiang, J. Xiong, J. Xu, W. Zhou, C. Liu, H. Shi, Q. Jiang, Effects of second dopants on electrical conductivity and thermopower of poly(3,4-ethylenedioxythiophene):poly(styrenesulfonate)-filled carbon black. Mater. Chem. Phys. **153**, 285–290 (2015)
100. H. Ju, M. Kim, J. Kim, Enhanced thermoelectric performance of highly conductive poly(3,4-ethylenedioxythiophene)/carbon black nanocomposites for energy harvesting. Microelectron. Eng. **136**, 8–14 (2015)
101. M. Culebras, C.M. Gómez, A. Cantarero, Thermoelectric measurements of PEDOT:PSS/expanded graphite composites. J. Mater. Sci. **48**(7), 2855–2860 (2012)
102. D. Yoo, J. Kim, S.H. Lee, W. Cho, H.H. Choi, F.S. Kim, J.H. Kim, Effects of one- and two-dimensional carbon hybridization of PEDOT:PSS on the power factor of polymer thermoelectric energy conversion devices. J. Mater. Chem. A **3**(12), 6526–6533 (2015)
103. X. Li, L. Liang, M. Yang, G. Chen, C.-Y. Guo, Poly(3,4-ethylenedioxythiophene)/graphene/carbon nanotube ternary composites with improved thermoelectric performance. Org. Electron. **38**, 200–204 (2016)
104. N. Toshima, N. Jiravanichanun, H. Marutani, Organic thermoelectric materials composed of conducting polymers and metal nanoparticles. J. Electron. Mater. **41**(6), 1735–1742 (2012)
105. N. Toshima, N. Jiravanichanun, Improvement of thermoelectric properties of PEDOT/PSS films by addition of gold nanoparticles: enhancement of seebeck coefficient. J. Electron. Mater. **42**(7), 1882–1887 (2013)
106. A. Yoshida, N. Toshima, Gold nanoparticle and gold nanorod embedded PEDOT:PSS thin films as organic thermoelectric materials. J. Electron. Mater. **43**(6), 1492–1497 (2013)
107. Y. Wang, K. Cai, S. Chen, S. Shen, X. Yao, One-step interfacial synthesis and thermoelectric properties of Ag/Cu-poly(3,4-ethylenedioxythiophene) nanostructured composites. J. Nanopart. Res. **16**(8), 2531 (2014)
108. Y. Liu, Z. Song, Q. Zhang, Z. Zhou, Y. Tang, L. Wang, J. Zhu, W. Luo, W. Jiang, Preparation of bulk AgNWs/PEDOT:PSS composites: a new model towards high-performance bulk organic thermoelectric materials. RSC Adv. **5**(56), 45106–45112 (2015)
109. A. Yoshida, N. Toshima, Thermoelectric properties of hybrid thin films of PEDOT-PSS and silver nanowires. J. Electron. Mater. **45**(6), 2914–2919 (2016)
110. X. Sun, Y. Wei, J. Li, J. Zhao, L. Zhao, Q. Li, Ultralight conducting PEDOT:PSS/carbon nanotube aerogels doped with silver for thermoelectric materials. Sci.China Mater. **60**(2), 159–166 (2017)
111. H. Song, C. Liu, H. Zhu, F. Kong, B. Lu, J. Xu, J. Wang, F. Zhao, Improved thermoelectric performance of free-standing PEDOT:PSS/Bi2Te3 films with low thermal conductivity. J. Electron. Mater. **42**(6), 1268–1274 (2013)
112. Y. Du, K.F. Cai, S. Chen, P. Cizek, T. Lin, Facile preparation and thermoelectric properties of Bi(2)Te(3) based alloy nanosheet/PEDOT:PSS composite films. ACS Appl. Mater. Interfaces **6**(8), 5735–5743 (2014)
113. K. Wei, T. Stedman, Z.-H. Ge, L.M. Woods, G.S. Nolas, A synthetic approach for enhanced thermoelectric properties of PEDOT:PSS bulk composites. Appl. Phys. Lett. **107**(15), 153301 (2015)
114. J. Xiong, L. Wang, J. Xu, C. Liu, W. Zhou, H. Shi, Q. Jiang, F. Jiang, Thermoelectric performance of PEDOT:PSS/Bi2Te3-nanowires: a comparison of hybrid types. J. Mater. Sci. Mater. Electron. **27**(2), 1769–1776 (2015)

115. Y. Wang, K. Cai, X. Yao, Facile fabrication and thermoelectric properties of PbTe-modified poly(3,4-ethylenedioxythiophene) nanotubes. ACS Appl. Mater. Interfaces **3**(4), 1163–1166 (2011)

116. Y.Y. Wang, K.F. Cai, X. Yao, One-pot fabrication and enhanced thermoelectric properties of poly(3,4-ethylenedioxythiophene)-Bi2S3 nanocomposites. J. Nanopart. Res. **14**(5), 848 (2012)

117. W. Zheng, P. Bi, H. Kang, W. Wei, F. Liu, J. Shi, L. Peng, Z. Wang, R. Xiong, Low thermal conductivity and high thermoelectric figure of merit in p-type Sb2Te3/poly(3,4-ethylenedioxythiophene) thermoelectric composites. Appl. Phys. Lett. **105**(2), 023901 (2014)

118. S.W. Finefrock, X. Zhu, Y. Sun, Y. Wu, Flexible prototype thermoelectric devices based on Ag(2)Te and PEDOT:PSS coated nylon fibre. Nanoscale **7**(13), 5598–5602 (2015)

119. E.W. Zaia, A. Sahu, P. Zhou, M.P. Gordon, J.D. Forster, S. Aloni, Y.S. Liu, J. Guo, J.J. Urban, Carrier scattering at alloy nanointerfaces enhances power factor in PEDOT:PSS hybrid thermoelectrics. Nano Lett. **16**(5), 3352–3359 (2016)

120. E.J. Bae, Y.H. Kang, K.S. Jang, C. Lee, S.Y. Cho, Solution synthesis of telluride-based nano-barbell structures coated with PEDOT:PSS for spray-printed thermoelectric generators. Nanoscale **8**(21), 10885–10890 (2016)

121. C. Liu, F. Jiang, M. Huang, B. Lu, R. Yue, J. Xu, Free-standing PEDOT-PSS/Ca3Co4O9 composite films as novel thermoelectric materials. J. Electron. Mater. **40**(5), 948–952 (2010)

122. X. Wang, F. Meng, H. Tang, Z. Gao, S. Li, S. Jin, Q. Jiang, F. Jiang, J. Xu, Design and fabrication of low resistance palm-power generator based on flexible thermoelectric composite film. Synth. Met. **235**, 42–48 (2018)

123. A. Dey, A. Maity, M.A. Shafeeuulla Khan, A.K. Sikder, S. Chattopadhyay, PVAc/PEDOT:PSS/graphene–iron oxide nanocomposite (GINC): an efficient thermoelectric material. RSC Adv. **6**(27), 22453–22460 (2016)

124. D. Lee, S.Y. Sayed, S. Lee, C.A. Kuryak, J. Zhou, G. Chen, Y. Shao-Horn, Quantitative analyses of enhanced thermoelectric properties of modulation-doped PEDOT:PSS/undoped Si (001) nanoscale heterostructures. Nanoscale **8**(47), 19754–19760 (2016)

125. G.O. Park, J.W. Roh, J. Kim, K.Y. Lee, B. Jang, K.H. Lee, W. Lee, Enhanced thermoelectric properties of germanium powder/poly(3,4-ethylenedioxythiophene):poly(4-styrenesulfonate) composites. Thin Solid Films **566**, 14–18 (2014)

126. N.E. Coates, S.K. Yee, B. McCulloch, K.C. See, A. Majumdar, R.A. Segalman, J.J. Urban, Effect of interfacial properties on polymer-nanocrystal thermoelectric transport. Adv. Mater. **25**(11), 1629–1633 (2013)

127. S.K. Yee, N.E. Coates, A. Majumdar, J.J. Urban, R.A. Segalman, Thermoelectric power factor optimization in PEDOT:PSS tellurium nanowire hybrid composites. Phys. Chem. Chem. Phys. **15**(11), 4024–4032 (2013)

128. K.C. See, J.P. Feser, C.E. Chen, A. Majumdar, J.J. Urban, R.A. Segalman, Water-processable polymer-nanocrystal hybrids for thermoelectrics. Nano Lett. **10**(11), 4664–4667 (2010)

129. J.N. Heyman, B.A. Alebachew, Z.S. Kaminski, M.D. Nguyen, N.E. Coates, J.J. Urban, Terahertz and infrared transmission of an organic/inorganic hybrid thermoelectric material. Appl. Phys. Lett. **104**(14), 141912 (2014)

130. Q. Jiang, C. Liu, B. Lu, J. Xu, H. Song, H. Shi, D. Mo, Z. Wang, F. Jiang, Z. Zhu, PEDOT:PSS film: a novel flexible organic electrode for facile electrodeposition of dendritic tellurium nanostructures. J. Mater. Sci. **50**(14), 4813–4821 (2015)

131. H. Song, K. Cai, S. Shen, Enhanced thermoelectric properties of PEDOT/PSS/Te composite films treated with H2SO4. J. Nanopart. Res. **18**(12), 386 (2016)

132. E.J. Bae, Y.H. Kang, K.S. Jang, S.Y. Cho, Enhancement of thermoelectric properties of PEDOT:PSS and tellurium-PEDOT:PSS hybrid composites by simple chemical treatment. Sci. Rep. **6**, 18805 (2016)

133. H. Song, K. Cai, Preparation and properties of PEDOT:PSS/Te nanorod composite films for flexible thermoelectric power generator. Energy **125**, 519–525 (2017)

134. E.J. Bae, Y.H. Kang, C. Lee, S.Y. Cho, Engineered nanocarbon mixing for enhancing the thermoelectric properties of a telluride-PEDOT:PSS nanocomposite. J. Mater. Chem. A **5**(34), 17867–17873 (2017)

135. C. Li, F. Jiang, C. Liu, W. Wang, X. Li, T. Wang, J. Xu, A simple thermoelectric device based on inorganic/organic composite thin film for energy harvesting. Chem. Eng. J. **320**, 201–210 (2017)

136. F. Jiang, J. Xiong, W. Zhou, C. Liu, L. Wang, F. Zhao, H. Liu, J. Xu, Use of organic solvent-assisted exfoliated MoS2 for optimizing the thermoelectric performance of flexible PEDOT:PSS thin films. J. Mater. Chem. A **4**(14), 5265–5273 (2016)

137. A. Dey, S. Hadavale, M.A. Khan, P. More, P.K. Khanna, A.K. Sikder, S. Chattopadhyay, Polymer based graphene/titanium dioxide nanocomposite (GTNC): an emerging and efficient thermoelectric material. Dalton Trans. **44**(44), 19248–19255 (2015)

138. Y. Du, K. Cai, S.Z. Shen, W. Yang, J. Xu, T. Lin, ZnO flower/PEDOT:PSS thermoelectric composite films. J. Mater. Sci. Mater. Electron. **27**(10), 10289–10293 (2016)

139. J. Guo, H. Guo, D.S.B. Heidary, S. Funahashi, C.A. Randall, Semiconducting properties of cold sintered V 2 O 5 ceramics and Co-sintered V 2 O 5-PEDOT:PSS composites. J. Eur. Ceram. Soc. **37**(4), 1529–1534 (2017)

140. S. Ferhat, C. Domain, J. Vidal, D. Noël, B. Ratier, B. Lucas, Organic thermoelectric devices based on a stable n-type nanocomposite printed on paper. Sustain.Energy Fuels **2**(1), 199–208 (2018)

141. Y. Shiraishi, S. Hata, Y. Okawauchi, K. Oshima, H. Anno, N. Toshima, Improved thermoelectric behavior of poly(3,4-ethylenedioxythiophene)-poly(styrenesulfonate) using poly(N-vinyl-2-pyrrolidone)-coated GeO2 nanoparticles. Chem. Lett. **46**(7), 933–936 (2017)

142. X. Wang, F. Meng, H. Tang, Z. Gao, S. Li, F. Jiang, J. Xu, An effective dual-solvent treatment for improving the thermoelectric property of PEDOT:PSS with white graphene. J. Mater. Sci. **52**(16), 9806–9818 (2017)

143. T. Wang, C. Liu, X. Wang, X. Li, F. Jiang, C. Li, J. Hou, J. Xu, Highly enhanced thermoelectric performance of WS2 nanosheets upon embedding PEDOT:PSS. J. Polym. Sci. B Polym. Phys. **55**(13), 997–1004 (2017)

144. X. Li, C. Liu, T. Wang, W. Wang, X. Wang, Q. Jiang, F. Jiang, J. Xu, Preparation of 2D MoSe2/PEDOT:PSS composite and its thermoelectric properties. Mater. Res. Express **4**(11), 116410 (2017)

145. X. Cheng, L. Wang, X. Wang, G. Chen, Flexible films of poly(3,4-ethylenedioxythiophene):Poly(styrenesulfonate)/SnS nanobelt thermoelectric composites. Compos. Sci. Technol. **155**, 247–251 (2018)

146. Z.H. Ge, Y. Chang, F. Li, J. Luo, P. Fan, Improved thermoelectric properties of PEDOT:PSS polymer bulk prepared using spark plasma sintering. Chem. Commun. (Camb.) **54**(19), 2429–2431 (2018)

147. H.J. Lee, G. Anoop, H.J. Lee, C. Kim, J.-W. Park, J. Choi, H. Kim, Y.-J. Kim, E. Lee, S.-G. Lee, Y.-M. Kim, J.-H. Lee, J.Y. Jo, Enhanced thermoelectric performance of PEDOT:PSS/PANI–CSA polymer multilayer structures. Energy Environ. Sci. **9**(9), 2806–2811 (2016)

148. F. Jiang, L. Wang, C. Li, X. Wang, Y. Hu, H. Liu, H. Yang, F. Zhao, J. Xu, Effects of solvents on thermoelectric performance of PANi/PEDOT/PSS composite films. J. Polym. Res. **24**(5), 68 (2017)

149. X.Y. Wang, C.Y. Liu, L. Miao, J. Gao, Y. Chen, Improving the thermoelectric properties of polyaniline by introducing poly(3,4-ethylenedioxythiophene). J. Electron. Mater. **45**(3), 1813–1820 (2015)

150. H. Shi, C. Liu, Q. Jiang, J. Xu, B. Lu, F. Jiang, Z. Zhu, Three novel electrochemical electrodes for the fabrication of conducting polymer/SWCNTs layered nanostructures and their thermoelectric performance. Nanotechnology **26**(24), 245401 (2015)

151. C. Liu, J. Xu, B. Lu, R. Yue, F. Kong, Simultaneous increases in electrical conductivity and seebeck coefficient of PEDOT:PSS films by adding ionic liquids into a polymer solution. J. Electron. Mater. **41**(4), 639–645 (2012)

152. J. Luo, D. Billep, T. Waechtler, T. Otto, M. Toader, O. Gordan, E. Sheremet, J. Martin, M. Hietschold, D.R.T. Zahn, T. Gessner, Enhancement of the thermoelectric properties of PEDOT:PSS thin films by post-treatment. J. Mater. Chem. A **1**(26), 7576 (2013)

153. Y. Jia, X. Li, F. Jiang, C. Li, T. Wang, Q. Jiang, J. Hou, J. Xu, Effects of additives and post-treatment on the thermoelectric performance of vapor-phase polymerized PEDOT films. J. Polym. Sci. B Polym. Phys. **55**(23), 1738–1744 (2017)

154. T.-C. Tsai, H.-C. Chang, C.-H. Chen, W.-T. Whang, Widely variable Seebeck coefficient and enhanced thermoelectric power of PEDOT:PSS films by blending thermal decomposable ammonium formate. Org. Electron. **12**(12), 2159–2164 (2011)

155. F. Kong, C. Liu, J. Xu, Y. Huang, J. Wang, Z. Sun, Thermoelectric performance enhancement of poly(3,4-ethylenedioxythiophene):poly(styrenesulfonate) composite films by addition of dimethyl sulfoxide and urea. J. Electron. Mater. **41**(9), 2431–2438 (2012)

156. Q. Jiang, C. Liu, J. Xu, B. Lu, H. Song, H. Shi, Y. Yao, L. Zhang, Paper: an effective substrate for the enhancement of thermoelectric properties in PEDOT:PSS. J. Polym. Sci. B Polym. Phys. **52**(11), 737–742 (2014)

157. J. Li, Y. Du, R. Jia, J. Xu, S.Z. Shen, Thermoelectric properties of flexible PEDOT:PSS/polypyrrole/paper nanocomposite films. Materials **10**(7), 780 (2017)

158. Y. Du, J. Xu, Y. Wang, T. Lin, Thermoelectric properties of graphite-PEDOT:PSS coated flexible polyester fabrics. J. Mater. Sci. Mater. Electron. **28**(8), 5796–5801 (2016)

159. E.P. Tomlinson, M.J. Willmore, X. Zhu, S.W. Hilsmier, B.W. Boudouris, Tuning the thermoelectric properties of a conducting polymer through blending with open-shell molecular dopants. ACS Appl. Mater. Interfaces **7**(33), 18195–18200 (2015)

160. H. Zhou, W. Yao, G. Li, J. Wang, Y. Lu, Graphene/poly(3,4-ethylenedioxythiophene) hydrogel with excellent mechanical performance and high conductivity. Carbon **59**, 495–502 (2013)

161. K. Xu, G. Chen, D. Qiu, Convenient construction of poly(3,4-ethylenedioxythiophene)–graphene pie-like structure with enhanced thermoelectric performance. J. Mater. Chem. A **1**(40), 12395 (2013)

162. B.N. Reddy, A. Pathania, S. Rana, A.K. Srivastava, M. Deepa, Plasmonic and conductive Cu fibers in poly (3,4-ethylenedioxythiophene)/Cu hybrid films: Enhanced electroactivity and electrochromism. Sol. Energy Mater. Sol. Cells **121**, 69–79 (2014)

163. M.S. Cho, S.Y. Kim, J.D. Nam, Y. Lee, Preparation of PEDOT/Cu composite film by in situ redox reaction between EDOT and copper(II) chloride. Synth. Met. **158**(21–24), 865–869 (2008)

164. W. Shi, S. Qu, H. Chen, Y. Chen, Q. Yao, L. Chen, One-step synthesis and enhanced thermoelectric properties of polymer-quantum dot composite films. Angew. Chem. Int. Ed. Engl. **57**(27), 8037–8042 (2018)

165. W. Shi, T. Zhao, J. Xi, D. Wang, Z. Shuai, Unravelling doping effects on PEDOT at the molecular level: from geometry to thermoelectric transport properties. J. Am. Chem. Soc. **137**(40), 12929–12938 (2015)

Chapter 5
Electric Field Thermopower Modulation of 2D Electron Systems

Hiromichi Ohta

5.1 Introduction

Thermoelectric energy conversion technology attracts a great deal of attention for converting waste heat into electricity [1, 2]. The principle of thermoelectric energy conversion was discovered by Seebeck in 1821: a thermo-electromotive force (ΔV) is generated between two ends of a metal bar by introducing a temperature difference (ΔT) in the bar [3]. The value of $\Delta V / \Delta T$ is so-called thermopower or Seebeck coefficient (S), which is an important physical parameter to obtain high dimensionless thermoelectric figure of merit, $ZT = S^2 \cdot \sigma \cdot T \cdot \kappa^{-1}$, where Z, T, σ, and κ are the thermoelectric figure of merit, the absolute temperature, the electrical conductivity, and the thermal conductivity, respectively.

Currently, there are two frontiers in thermoelectric materials research: one is to reduce κ value using nanostructuring technique. Several high ZT materials were realized one after another [4–8]. The other one is to enhance $|S|$ value without reduction of σ by electronic density of states (DOS) modification [9–12]. Thus, the ZT value of thermoelectric materials can be dramatically enhanced by modifying the DOS in low-dimensional structures such as two-dimensional quantum wells, due to that an enhancement of $|S|$ occurs. Hicks and Dresselhaus [10] theoretically predicted that using superlattices can dramatically enhance the two-dimensional thermoelectric figure of merit, $Z_{2D}T$ of a quantum well for thermoelectric semi-conductors because only the $|S|$ value increases with the electronic DOS of the quantum well, while σ and κ remain constant [10]. In 2007, we observed an unusually large enhancement of $|S|$ value of $SrTiO_3$, which is five times larger than that of bulk, in two-dimensional electron gases (2DEGs), which are confined in

H. Ohta (✉)
Research Institute for Electronic Science, Hokkaido University, Sapporo, Japan
e-mail: hiromichi.ohta@es.hokudai.ac.jp

© Springer Nature Switzerland AG 2019
P. Mele et al. (eds.), *Thermoelectric Thin Films*,
https://doi.org/10.1007/978-3-030-20043-5_5

Nb:SrTiO$_3$/SrTiO$_3$ superlattices [13, 14] and/or TiO$_2$/SrTiO$_3$ heterointerfaces [14]. Thus, 2DEGs showing unusually large $|S|$ have attracted attention as a potential approach for developing high-performance thermoelectric materials.

Basically, the $|S|$ value of a thermoelectric material can be expressed by the following Mott equation [15]:

$$S = \frac{\pi^2}{3} \frac{k_B^2 T}{e} \left\{ \frac{d[\ln(\sigma(E))]}{dE} \right\}_{E=E_F},$$
$$= \frac{\pi^2}{3} \frac{k_B^2 T}{e} \left\{ \frac{1}{n} \cdot \frac{dn(E)}{dE} + \frac{1}{\mu} \cdot \frac{d\mu(E)}{dE} \right\}_{E=E_F},$$

where k_B, e, n, and μ are the Boltzmann constant, electron charge, carrier concentration, and carrier mobility, respectively. The $S^2 \cdot \sigma$ value of a thermoelectric material must be maximized by means of n because of the commonly observed trade-off relationship between S and σ in terms of n: σ increases almost linearly with increasing n, while $|S|$ decreases with n. Consequently, a great deal of effort in material engineering is needed to maximize $S^2 \cdot \sigma$ value. In addition, $|S|$ strongly depends on the energy derivative of the DOS at around the Fermi energy (E_F), $\left[\frac{\partial DOS(E)}{\partial E} \right]_{E=E_F}$. Thus, S is a good measure to characterize the material's electronic structure. However, the S measurements take very long time because many samples with different carrier concentration should be prepared. In order to overcome this difficulty, I have developed the electric field thermopower modulation method. Here I review the points of our electric field modulation measurement of S for 2DEGs in the several semiconductors including SrTiO$_3$, BaSnO$_3$, and AlGaN/GaN heterointerfaces.

5.2 Electric Field Thermopower Modulation Method [16–22]

The electric field thermopower modulation method was firstly demonstrated by Pernstich et al. [16]. They successfully measured the field modulated S of several organic semiconductors such as rubrene and pentacene by using the electric field thermopower modulation method. Figure 5.1 shows the measurement setup of the electric field modulated thermopower. For the S measurements, we used two Peltier devices, which were placed under the transistor, to generate a temperature difference between the source and the drain electrodes. Two thermocouples (K-type), which were mechanically attached at both edges of the channel, monitored the temperature difference (ΔT). The thermo-electromotive force (ΔV) and ΔT values were simultaneously measured at room temperature, and the slope of the ΔV–ΔT plots yielded the S values.

In order to characterize a thermoelectric material, the power factor $S^2 \cdot \sigma$ value must be maximized by measuring S and σ with varied n. Although many specimens with varied carrier concentration are required to clarify the relationship between the S and the n_{3D} (Fig. 5.2a), only one good specimen with three-terminal TFT structure is required for the electric field thermopower modulation method (Fig. 5.2b) [19].

Fig. 5.1 Measurement setup of the electric field modulated thermopower. (**a**) The thin-film transistor sample is put on the gap between two Peltier devices. One Peltier device is a heater and the other one is a cooler when the electric current is applied simultaneously. (**b**) Magnified photograph. Two tiny thermocouples (K-type, 25 μm) are used. The typical channel length is 800 μm and the width is 400 μm. When the gate voltage is applied, the thermopower of the channel is modulated. The thermos-electromotive force is measured using the source and drain electrodes

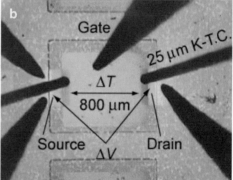

By applying the gate voltage, the carriers can be accumulated at the interface between the gate insulator and the thermoelectric material. Thus, the relationship between the sheet carrier concentration (n_{2D}) and S can be measured quickly by using the electric field thermopower modulation method.

There are several advantages in our electric field thermopower modulation method (Figs. 5.3, 5.4, and 5.5). First, the channel thickness can be controlled together with the sheet carrier concentration by modulating the gate voltage (V_g) (Fig. 5.3). Although artificial superlattice thermoelectric materials should exhibit an enhanced $|S|$, fabrication of artificial superlattice structures using state-of-the-art thermoelectric materials is extremely difficult due to their complicated crystal structures. Furthermore, the production cost of such superlattice materials, which

Fig. 5.2 Comparison between the conventional thermoelectric measurement and the electric thermopower modulation measurement of a material system. (**a**) Conventional thermoelectric measurement. Many specimens with varied carrier concentration are required to clarify the relationship between the thermopower and the carrier concentration. (**b**) Electric field thermopower modulation measurement. Only one good specimen with three-terminal TFT structure is required. The measurement finishes quickly. (Reprinted from [19] with permission from Springer Nature)

a   Many specimen with varied carrier conc.

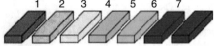

Each measurement takes very long time.

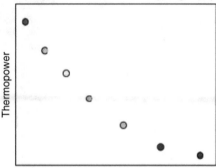

3D carrier concentration, n_{3D} (log scale)

b   Only one "good" specimen with three-terminal TFT structure

The measurement finishes quickly.

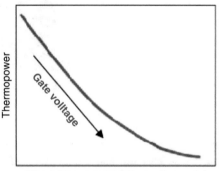

2D carrier concentration, n_{2D} (log scale)

can be fabricated by precise vacuum deposition methods such as molecular beam epitaxy (MBE) and pulsed laser deposition (PLD) at high temperatures, is extremely high. By using the electric field thermopower modulation method, the channel thickness may be reduced by the strong electric field application. One can expect that the channel thickness becomes thinner from 3D density of states to 2D density of states when increasing the V_g. As shown in Fig. 5.4, the n_{2D} increases with the gate voltage, while the effective thickness (t_{eff}) decreases. Therefore, the S decreases

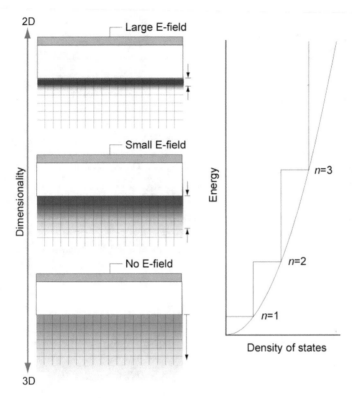

Fig. 5.3 Advantage of the electric field thermopower modulation measurement (1). When increasing the gate voltage, the channel thickness becomes thinner from 3D density of states to 2D density of states

with the V_g first. But it may show V-shaped up turn when the t_{eff} is thinner than the de Broglie wavelength (λ_D) due to that the electronic DOS change from 3D to 2D as described above.

Furthermore, the electric field-induced 2DEG would show very high thermoelectric power factor (PF) because of its very high mobility (Fig. 5.5). In the case of conventional impurity doped semiconductor, the mobility decreases with increasing n due to ionized impurity scattering. On the other hand, there is no impurity in the electric field accumulated 2DEG channel. Therefore, high mobility can be maintained in the case of the electric field accumulated 2DEG channel.

5.3   Electric Field Thermopower Modulation of SrTiO₃ [22]

In order to test the ability of the electric field thermopower modulation method, I used SrTiO₃ single crystal-based thin-film transistor structure (Fig. 5.6). Electron-doped SrTiO₃ is known as a candidate of an n-type oxide thermoelectric material

Fig. 5.4 Advantage of the electric field thermopower modulation measurement (2). (**a**) Schematic illustration of a filed effect transistor when very high gate voltage is applied. (**b**) Relationship between the thermopower, sheet carrier concentration, and effective thickness of the channel as a function of the electric field. One can expect that V-shaped up turn of thermopower occurs when very high electric field is applied. (Reprinted from [20] with permission from John Wiley and Sons)

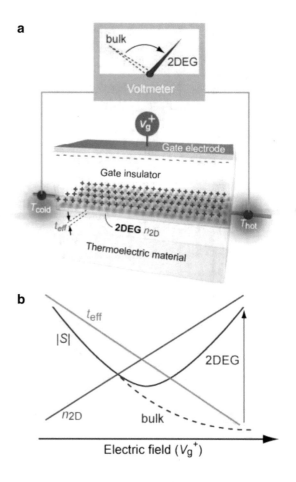

[14, 23–27]. First, 20-nm-thick metallic Ti films, used as the source and drain electrodes, were deposited through a stencil mask by electron beam (EB) evaporation (base pressure $\sim 10^{-4}$ Pa, no substrate heating) onto a stepped $SrTiO_3$ substrate (10 mm $\times$ 10 mm $\times$ 0.5 mm), treated with NH_4-buffered HF (BHF) solution [28]. Second, 150-nm-thick amorphous C12A7 ($12CaO \cdot 7Al_2O_3$) film was deposited through a stencil mask by PLD (KrF excimer laser, fluence ~ 3 J cm^{-2} pulse^{-1}, oxygen pressure ~ 0.1 Pa) using dense polycrystalline C12A7 ceramic as target. Finally, 20-nm-thick metallic Ti films, used as the gate electrode, was deposited through a stencil mask by EB evaporation. After the deposition processes, the devices were annealed at 200 °C for 30 min in air to reduce the off current.

Halo peaking only at $2\theta \sim 30°$ was observed in the glancing incidence X-ray diffraction pattern of the resulting C12A7 film (Fig. 5.7a). This pattern was very similar to the halo pattern of the C12A7 glass. The density of the C12A7 film was ~ 2.9 g cm^{-3} as evaluated by the X-ray reflectivity and was in a good correspondence with that of C12A7 glass (2.92 g cm^{-3}) [29]. It should be noted

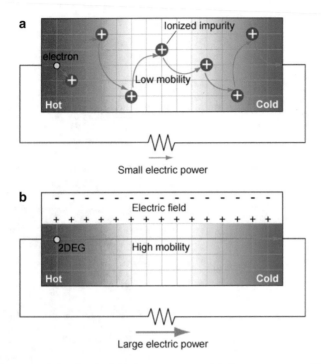

Fig. 5.5 Advantage of the electric field thermopower modulation measurement (3). Schematic illustration of thermoelectric power generation in an n-type semiconductor. (**a**) Conventional impurity doped n-type semiconductor. Carrier electron flow from the hot to the cold side due to the Seebeck effect. Low mobility of the carrier electron originates from ionized impurity scattering. (**b**) Electric field-induced high-mobility 2DEG. Larger electric power can be obtained compared with (**a**). (Reprinted from Ref. [17] with permission from John Wiley and Sons)

that any grain structure in the as-deposited film was not observed in the topographic AFM image (Fig. 5.7a, inset). Furthermore, no grain structure was observed in the cross sectional high-resolution transmission electron microscope image of the a-C12A7/SrTiO$_3$ interface region (Fig. 5.7b). A broad halo pattern is seen in the selected area electron diffraction patterns of a-C12A7.

Figure 5.8 shows typical (a) transfer and (b) output characteristics of the resultant FET. Drain current (I_d) of the TFT increased markedly as the V_g increased, hence the channel was n-type, and electron carriers were accumulated by positive V_g (Fig. 5.8a). A small hysteresis ($\sim$0.5 V) in I_d, probably due to traps at the a-C12A7/SrTiO$_3$ interface, was also seen (Fig. 5.8a). We observed a clear pinch-off and saturation in I_d (Fig. 5.8b), indicating that the operation of this TFT conformed to standard FET theory. The on-to-off current ratio, sub-threshold swing, and threshold gate voltage (V_{gth}), which were obtained from a linear fit of an $I_d^{0.5} - V_g$ plot, are >10^6, $\sim$0.3 V decade^{-1} and +1.1 V, respectively. We noted the dramatic increase in effective mobility (μ_{eff}) of the TFT increases drastically with V_g and

Fig. 5.6 Three-terminal
thin-film transistor structure
on a SrTiO₃ single crystal
plate. (**a**) Schematic device
structure and (**b**) photograph
of the SrTiO₃-TFT. Ti films
(20-nm thick) are used as the
source, drain, and gate
electrodes. A 150-nm-thick
amorphous 12CaO·7Al₂O₃
(C12A7) film is used as the
gate insulator. Channel length
(L) and channel width (W) are
200 and 400 μm,
respectively. (Reprinted from
[22] with permission from
AIP Publishing)

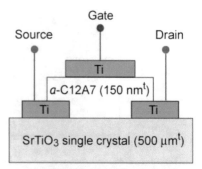

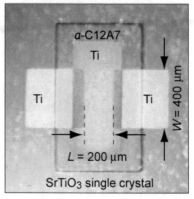

reaches at ∼2 cm² V⁻¹ s⁻¹, which is ∼30% of the room temperature Hall mobility
(μ_{Hall}) of electron-doped SrTiO₃ (μ_{Hall} ∼6 cm² V⁻¹ s⁻¹) [24] (Fig. 5.8a, inset).

Then we measured electric field modulated thermopower of the SrTiO₃-TFT at
room temperature (Fig. 5.9). First, a temperature difference ($\Delta T = 0.2$–1.5 K) was
introduced between the source and drain electrodes by using two Peltier devices
(Fig. 5.9 left). Then, thermo-electromotive force (V_{TEMF}) was measured during the
V_g-sweeping. The values of S were obtained from the slope of V_{TEMF}–ΔT plots
(inset). Figure 5.9 (right) shows S–V_g plots for the SrTiO₃-TFT. The S values are
negative, confirming that the channel is n-type. The $|S|$ value gradually decreases
from 1600 to 580 μV K⁻¹, which corresponds to an increase of the n_{3D} up to
∼8 × 10¹⁸ cm⁻³ (Fig. 5.10), due to the fact that electron carriers are accumulated by
positive V_g (up to +30 V). The n_{2D} slightly exceeded 10¹³ cm⁻². Thus, the effective
thickness (n_{2D}/n_{3D}) can be estimated as ∼12 nm.

5.4   Electric Field Thermopower Modulation of BaSnO₃ [18]

In order to further clarify the ability of the electric field thermopower modulation
method, we tested BaSnO₃-based TFT, which is known as a transparent oxide
semiconductor (TOS). TOSs with a relatively high electrical conductivity and a

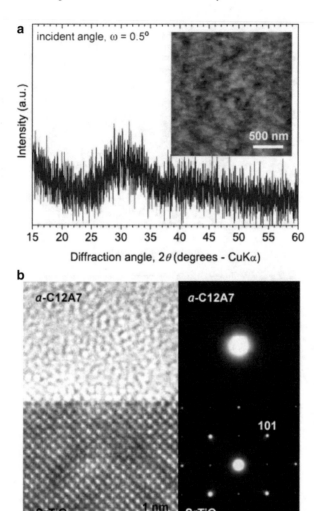

Fig. 5.7 Gate insulator, amorphous $12CaO \cdot 7Al_2O_3$ (C12A7) film. (**a**) Glancing angle XRD pattern of a 150-nm-thick a-C12A7 layer. Broad halo of a-C12A7 is seen around $2\theta \sim 30°$. Topographic AFM image (2×2 μm^2) of the C12A7 film on the $SrTiO_3$ surface is also shown in the inset. (**b**) Cross-sectional high-resolution transmission electron microscope (HRTEM) images of the 150-nm-thick a-C12A7/$SrTiO_3$ heterointerface, showing an abrupt interface of a-C12A7/$SrTiO_3$. Featureless image of a-C12A7 clearly indicates that the a-C12A7 layer is glass. A broad halo pattern is seen in the selected area electron diffraction patterns of a-C12A7 (right). (Reprinted from [22] with permission from AIP Publishing)

large bandgap ($E_g > 3.1$ eV) are commonly used as transparent electrodes and channel semiconductors for TFT-driven flat panel displays such as liquid crystal displays (LCDs) and organic light emitting diodes (OLEDs) [30]. TOSs materials include Sn-doped In_2O_3 (ITO) and $InGaZnO_4$-based oxides. Novel TOSs exhibiting

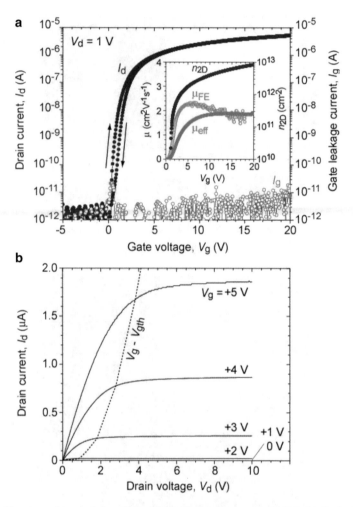

Fig. 5.8 Transistor characteristics of the SrTiO$_3$-based three-terminal TFT. (**a**) Typical transfer and (**b**) output characteristics of the SrTiO$_3$-TFT with 150-nm-thick a-C12A7 ($\varepsilon_r = 12$) gate insulator at room temperature. Channel length (L) and channel width (W) are 200 and 400 μm, respectively. Effective mobility (μ_{eff}), field-effect mobility (μ_{FE}) and sheet carrier concentration (n_{2D}) vs. V_g plots for the SrTiO$_3$-TFT are also shown in the inset of (**a**). The dotted line in (**b**) indicates V_g-V_{gth} value. (Reprinted from [22] with permission from AIP Publishing)

higher carrier mobilities have been intensively explored since the TFT performance strongly depends on the carrier mobility of the channel semiconductor. In 2012, Kim et al. (2012) reported that a La-doped BaSnO$_3$ (space group: $Pm\bar{3}m$, cubic perovskite structure, $a = 4.115$ Å, E_g ~3.1 eV) single crystal grown by the flux method exhibits a very high mobility (μ_{Hall} ~320 cm^2 V^{-1} s^{-1}) at room temperature [31, 32]. This report inspired the current interest of BaSnO$_3$ films and BaSnO$_3$-based TFTs [33–38].

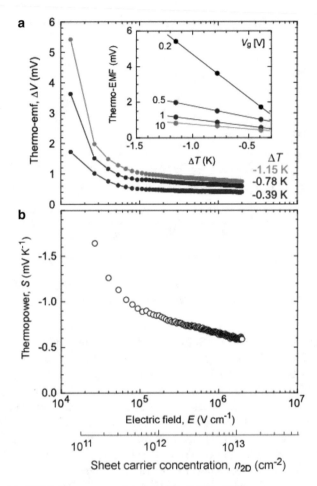

Fig. 5.9 Electric field thermopower modulation of the SrTiO$_3$-TFT at room temperature. (**a**) First, the temperature difference (ΔT) is added to the channel. Then, the thermo-electromotive force (EMF) is measured during the gate voltage sweeping. After repeating this measurement several times, the thermo-EMF is plotted as a function of ΔT (inset). The slope is the thermopower. (**b**) Thermopower as a function of the sheet carrier concentration (n_{2D}). This measurement finishes within an hour

Since the mobility is expressed as $\mu = e \cdot \tau \cdot m^{*-1}$, where e, τ, and m^* are the electron charge, carrier relaxation time, and carrier effective mass, respectively, the high mobility of the La-doped BaSnO$_3$ single crystal should be due to both a small m^* and a large τ. Generally, τ value of epitaxial films is smaller than that of the bulk single crystal due to the fact that the carrier electrons are scattered at dislocations, which are originated from the lattice mismatch (δ) and at other structural defects, in addition to optical phonon scattering. The estimated misfit dislocation spacing d is 7.4 nm because δ between BaSnO$_3$ and SrTiO$_3$ ($a = 3.905$ Å) is +5.3%.

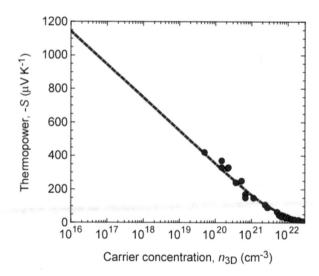

Fig. 5.10 Thermopower of electron-doped SrTiO$_3$ bulk at room temperature. Plots: Observed, Dotted line: Calculated. The slope of $|S| - \log n$ is -198 μV K^{-1}/decade

BaSnO$_3$ films grown on (001) SrTiO$_3$ substrates exhibit rather small mobilities (μ_{Hall} 26–100 cm^2 V^{-1} s^{-1}) [32, 36, 39] compared with those grown on (001) PrScO$_3$ ($a = 4.026$ Å, $\delta = +2.2\%$, $d \sim 17.7$ nm) (μ_{Hall} 150 cm^2 V^{-1} s^{-1}) [36]. On the other hand, m^* only depends on the electronic structure of the material. Many theoretical and experimental values of m^*, which were mostly determined from the optical properties, have been reported (e.g., theoretical values of $\sim$0.06 m_0 [40], $\sim$0.4 m_0 [32], and $\sim$0.2 m_0 [41], and experimental values of 3.7 m_0 [42], 0.61 m_0 [43], $\sim$0.35 m_0 [44], $\sim$0.396 m_0 [45], 0.27 $\pm$ 0.05 m_0 [39], 0.19 $\pm$ 0.01 m_0 [37]). Consequently, determining the intrinsic m^* value is almost impossible.

In order to clarify the m^* in La-doped BaSnO$_3$, we fabricated La-doped BaSnO$_3$ epitaxial films on (001)-oriented SrTiO$_3$ single crystal substrate by PLD method and measured the thermopower. Figure 5.11 shows the change in the S of La-doped BaSnO$_3$ films as a function of the n_{3D} at room temperature. Red line is to guide the eye. Calculated effective mass m^* values are also shown. Gray line is the $S - n_{3D}$ curve calculated using $m^* = 0.4 \pm 0.1$ m_0. Dotted line is the threshold of the degenerate/non-degenerate semiconductor around $n_{3D} = 1.4 \times 10^{19}$ cm^{-3}. Inset shows a schematic explanation of the energy dependence of the DOS. We have clarified the intrinsic m^* value of BaSnO$_3$, $m^* = 0.40 \pm 0.01$ m_0, by measuring the thermopower by the conventional way. However, it took long time for the sample preparation and the measurement.

In order to reduce the measurement time, we fabricated the three-terminal TFT structure on the BaSnO$_3$ film with same manner of that of SrTiO$_3$. Figure 5.12 summarizes the typical transistor characteristics such as the transfer and output characteristics, threshold voltage, field-effect mobility (μ_{FE}) of the resultant BaSnO$_3$-TFT. The drain current I_d increased markedly as gate voltage V_g increased

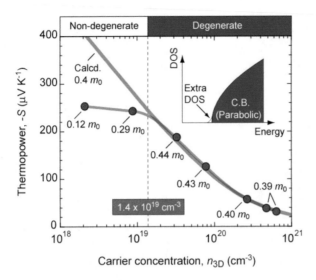

Fig. 5.11 Thermopower of La-doped BaSnO₃ at room temperature. Change in the thermopower S of La-doped BaSnO₃ films grown on (001) SrTiO₃ substrates as a function of the carrier concentration n_{3D} at room temperature. Red line is to guide the eye. Calculated effective mass m^* values are also shown. Gray line is the $S - n_{3D}$ curve calculated using $m^* = 0.4 \pm 0.1\ m_0$. Dotted line is the threshold of the degenerate/non-degenerate semiconductor around $n_{3D} = 1.4 \times 10^{19}$ cm^{-3}. Inset shows a schematic explanation of the energy dependence of the DOS. (Reprinted from [18] with permission from the American Physical Society)

(Fig. 5.12a), indicating that the channel was an n-type and the electron carriers accumulated by a positive V_g. An on–off current ratio of $\sim10^3$ was obtained for $V_d = 1$ V. A rather large V_{th} of +5.5 V was observed from the linear fit of the $I_d^{0.5} - V_g$ plot (Fig. 5.12b). A clear pinch-off behavior and current saturation in I_d revealed that the resultant TFT obeyed the standard field-effect transistor theory (Fig. 5.12c). The μ_{FE} drastically increased with V_g and became saturated at ~40 cm^2 V^{-1} s^{-1}, which was $\sim60\%$ of the room temperature μ_{Hall} of La-doped BaSnO₃ ($\mu_{Hall} \sim67$ cm^2 V^{-1} s^{-1}).

Figure 5.13 shows the electric field modulated S of the resultant BaSnO₃ TFT as a function of sheet carrier concentration n_{2D}. The $|S|$ value gradually decreased from 308 to 120 μV K^{-1} with n_{2D}, which is consistent with the La-doped BaSnO₃ films (Fig. 5.11). It should be noted that a deflection point occurred around $(|S|, n_{2D}) = (240\ \mu$V K$^{-1}, 1.8 \times 10^{12}$ cm$^{-2})$. An almost linear relationship with a slope with of ~200 μV K^{-1}/decade was observed in the $S - \log n_{2D}$ plot when n_{2D} exceeded 1.8×10^{12} cm^{-2}. Below 1.8×10^{12} cm^{-2}, the slope was not linear. This observation is similar to that of the La-doped BaSnO₃ films; the degenerate/non-degenerate threshold was around $(|S|, n_{2D}) = (240\ \mu$V K$^{-1}, 1.8 \times 10^{12}$ cm$^{-2})$. When the E_F locates above the parabolic-shaped conduction band bottom, rather high mobility was observed. On the contrary, very low carrier mobility was observed when the E_F lays below the threshold, most likely due to that the tail states suppress

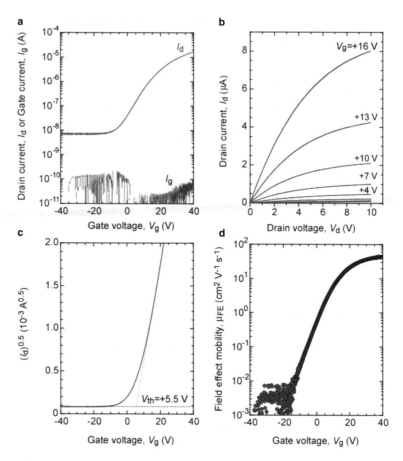

Fig. 5.12 Typical transistor characteristics of the BaSnO$_3$ TFT at room temperature. (**a**) Transfer characteristics $I_d - V_g$ at $V_d = 1$ V, (**b**) $I_d^{0.5} - V_g$ plot, (**c**) Output characteristics $I_d - V_d$, and (**d**) Field-effect mobility μ_{Hall}. On–off current ratio is $\sim 10^3$. Threshold voltage V_{th} is $+5.5$ V. Maximum field-effect mobility μ_{FE} is ~ 40 cm^2 V^{-1} s^{-1}. (Reprinted from [18] with permission from the American Physical Society)

the carrier mobility. It should be noted that the effective thickness (n_{2D}/n_{3D}) is only 1 nm, one order magnitude thinner than the SrTiO$_3$ case.

5.5 Unusually Large Thermopower Modulation in Water-Gated SrTiO$_3$ TFT [20, 21]

As described above, we verify a part of the advantage of the electric field thermopower modulation method. However, unusually large enhancement of $|S|$ (Fig. 5.4) has not been observed: the $|S|$ value decreased monotonically with electric

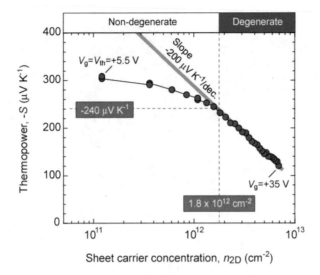

Fig. 5.13 Electric field modulated S of the resultant $BaSnO_3$ TFT as a function of the sheet carrier concentration at room temperature. Inflection point occurs around ($|S|$, n_{2D}) = (240 $\mu V\ K^{-1}$, $1.8 \times 10^{12}\ cm^{-2}$). Slope of $S - \log n_{2D}$ plot is ~200 $\mu V\ K^{-1}$/decade (gray line) when n_{2D} exceeds $1.8 \times 10^{12}\ cm^{-2}$. These data indicate that the threshold of a degenerate/non-degenerate semiconductor is around $n_{2D} = 1.8 \times 10^{12}\ cm^{-2}$. (Reprinted from [18] with permission from the American Physical Society)

field due to an increase in the n_{2D} value, most likely due to that the 2DEG was thicker (~12 nm) than the critical thickness ($\approx \lambda_D$).

To verify our hypothesis, we explored an excellent gate insulator, which can strongly accumulate electrons into an extremely thin 2DEG layer. Although liquid electrolytes including "gel" would be very useful to accumulate electrons at the 2DEG using their huge capacitance, they would not be suitable for the present application without sealing due to liquid leakage problem. In order to overcome this problem, we discovered that water-infiltrated nanoporous $12CaO\cdot7Al_2O_3$ glass (CAN) [21] works as an excellent gate insulator, and it can strongly accumulate electrons up to $n_{2D} \sim 10^{15}\ cm^{-2}$, which is two orders of magnitude higher than that accumulated by conventional gate insulators ($n_{2D} \sim 10^{13}\ cm^{-2}$). It should be noted that CAN is a chemically stable rigid glassy solid, showing excellent adhesion with oxide surface and no water leakage.

Figure 5.14a shows schematic illustration of a $SrTiO_3$-TFT using CAN as the gate insulator. The transfer characteristic curve of the TFT shows very large anti-clockwise hysteresis (Fig. 5.14b), indicating motion of the mobile ions in the gate insulator. Figure 5.14c, d shows the cross-sectional TEM images of the CAN-gated $SrTiO_3$-TFT coupled with the parallel schematic illustrations on the 2DEG-formation mechanism before (Fig. 5.14c, e) and after (Fig. 5.14d, f) applying positive V_g (+40 V). A trilayer structure composed of Ti/CAN/$SrTiO_3$ is observed in Fig. 5.14c. Thickness of the CAN layer at the virgin state is ~200 nm, while

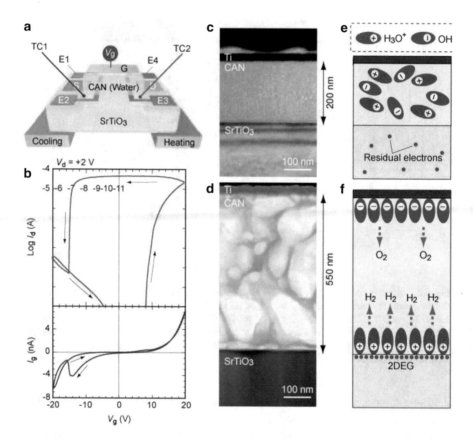

Fig. 5.14 Water electrolysis-induced charge carrier accumulation in SrTiO₃. (**a**) Schematic illustration of a SrTiO₃ TFT using CAN (C12A7 with nanopores). (**b**) Transfer characteristic curve of the TFT. Very large anti-clockwise hysteresis is seen. (**c, d**) Cross-sectional TEM images of the CAN-gated SrTiO₃ TFT coupled with the parallel schematic illustrations on the 2DEG-formation mechanism before (**c, e**) and after (**d, f**) applying positive V_g (+40 V). (**c**) A trilayer structure composed of Ti/CAN/SrTiO₃ is observed. Thickness of the CAN layer at the virgin state is ∼200 nm, while it is ∼550 nm (**f**), i.e., ∼2.8 times thicker than the virgin state, after applying the positive V_g. (Reprinted from [20] with permission from John Wiley and Sons)

it is ∼550 nm (d), i.e., ∼2.8 times thicker than the virgin state, after applying the positive V_g. This behavior can be schematically explained as shown in Fig. 5.14e, f.

The observed $|S|$ values were plotted as a function of n_{2D} (Fig. 5.15a) and t_{eff} (Fig. 5.15b) on a logarithmic scale. Additionally, the simulated bulk $|S|_{tot}$ values are plotted (gray dotted line). It should be noted that the $|S|$ values for the 2DEG in both the CAN- and the dense a-C12A7-gated TFTs were smoothly connected. $|S|$ initially decreased with n_{2D} from ∼1150 to ∼250 µV K⁻¹. Simultaneously, the t_{eff} decreased from ∼200 to ∼2 nm. In this region, the n_{2D} dependence of $|S|$ was similar to that of the simulated bulk values. When n_{2D} exceeded ∼2.5 × 10¹⁴ cm⁻², $|S|$ increased drastically to ∼950 µV K⁻¹, while t_{eff} remained nearly constant

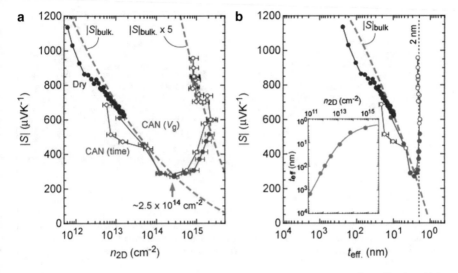

Fig. 5.15 Electric field modulation of thermopower–sheet carrier concentration–effective thickness relation. $|S|$ values of the 2DEGs are plotted as a function of (**a**) n_{2D} and (**b**) t_{eff} on a logarithmic scale. $|S|$ initially decreases with n_{sheet} from ~1150 to ~250 μV K^{-1}. Simultaneously, t_{eff} decreases from ~100 to ~2 nm. In this region, the n_{2D} dependence of $|S|$ is similar to that of simulated bulk values (gray dotted line). When the n_{2D} value exceeds ~2.5 × 10^{14} cm^{-2}, $|S|$ increases drastically and is modulated from ~600 to ~950 μV K^{-1}, while t_{eff} remains nearly constant (~2 nm). $|S|$ vs. log n_{2D} relation is approximately five times larger than that of the bulk, clearly indicating that the electric field-induced 2DEG in SrTiO$_3$ exhibits an unusually large $|S|$. (Reprinted from [20] with permission from John Wiley and Sons)

(~2 nm), demonstrating an unusually large $|S|$ for the electric field-induced 2DEG. The $|S|$ of the 2DEG in the CAN-gated FET was reversibly modulated from 600 (n_{2D} ~2 × 10^{15} cm^{-2}) to 950 μV K^{-1} (n_{2D} ~8 × 10^{14} cm^{-2}). The $|S|$ vs. log n_{2D} relation was approximately five times larger than that that of the bulk. Thus, we successfully demonstrated that an electric field-induced 2DEG yields unusually large enhancement of $|S|$. Moreover, because the present electric field thermopower modulation method is simple and effectively verifies the performance of thermoelectric materials, it may accelerate the development of nanostructures for high-performance thermoelectric materials.

5.6 Electric Field Thermopower Modulation of AlGaN/GaN Interfaces [17]

Finally, I would like to explain that a high-mobility two-dimensional electron gas (2DEG) at a semiconductor heterointerface is a viable solution to overcome the bottleneck in thermoelectric trade-off relations. In 2DEG, μ is not suppressed since the high-mobility channel lacks an impurity (Fig. 5.5b). Furthermore, the PF by the

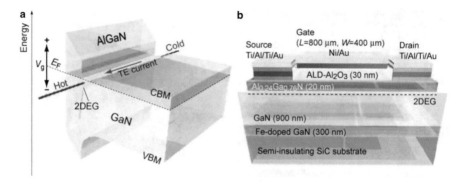

Fig. 5.16 Electric field thermopower modulation measurement of AlGaN/GaN MOS-HEMT. (**a**) Schematic energy band diagram of 2DEG confined at an AlGaN/GaN heterointerface. CBM and VBM denote the conduction band minimum and the valence band maximum, respectively. Thermoelectric properties can be modulated by applying a gate voltage (V_g). (**b**) Schematic illustration of the AlGaN/GaN MOS-HEMT. (Reprinted from [17] with permission from John Wiley and Sons)

electric field carrier concentration modulation can be optimized by the metal-oxide-semiconductor (MOS) structure on such a 2DEG.

In this study, we investigate the PF of a 2DEG, which is induced at an AlGaN/GaN heterointerface by the electric field thermopower modulation method (Fig. 5.16a). The maximized PF of the 2DEG is ~ 9 mW m^{-1} K^{-2} at room temperature, which is an order magnitude greater than that of the doped GaN bulk and two- to sixfold greater than those of the state-of-the-art thermoelectric materials, while maintaining a higher σ (=6030 S cm^{-1}) than the state-of-the-art thermoelectric materials ($\sigma = 1000$–2500 S cm^{-1}).

We fabricated an Al$_2$O$_3$/AlGaN/GaN metal-oxide-semiconductor high-electron-mobility transistor (MOS-HEMT) [46] and measured the thermoelectric properties of the 2DEG as schematically shown in Fig. 5.16b. We used a commercially available Al$_{0.24}$Ga$_{0.76}$N (20 nm)/GaN (900 nm) heterostructure film, which was grown on a semi-insulating (0001) SiC substrate by metal organic chemical vapor deposition. The sheet resistance (R_s), μ_{Hall}, and n_{2D} were 423 Ω sq.$^{-1}$, 1730 cm^2 V^{-1} s^{-1}, and 8.53 $\times$ 10^{12} cm^{-2}, respectively, at room temperature.

Figure 5.17 summarizes the carrier transport properties of the MOS-HEMT at room temperature. Applying a V_g from -9 V to $+4$ V at a V_d of $+10$ V dramatically modulates the I_d from 7 nA to 7 mA ($\equiv$on-to-off current ratio$\sim 10^6$) (Fig. 5.17a). The I_g is ~ 300 pA when V_g is less than $+2$ V, while the $I_d^{0.5}$ vs. V_g plot indicates that the V_{th} is -7.98 V. The output characteristic curves clearly show the pinch-off behavior and the current saturation of I_d (Fig. 5.17b), indicating that the characteristic of the MOS-HEMT obeys the standard transistor theory. The n_{2D} value is 8.32 $\times$ 10^{12} cm^{-2}, which agrees well with that obtained from the Hall measurement ($n_{2D} = 8.53 \times 10^{12}$ cm^{-2}). In the present MOS-HEMT, n_{2D} can be modulated from $\sim 10^{11}$ cm^{-2} up to 1.25 $\times$ 10^{13} cm^{-2}.

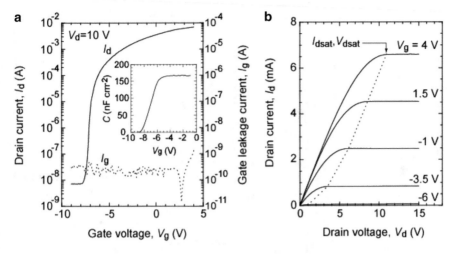

Fig. 5.17 Carrier transport properties of 2DEG in AlGaN/GaN MOS-HEMT at room temperature. (a) Transfer $I_d - V_g$ characteristic at $V_d = 10$ V. $I_g - V_g$ characteristic is also plotted. Inset shows the $C_i - V_g$ characteristic. (b) Output $I_d - V_d$ characteristics (-6 V $\leq V_g \leq +4$ V). Pinch-off and current saturation behaviors are clearly observed. (Reprinted from [17] with permission from John Wiley and Sons)

The observed S values are always negative, indicating that the channel is an n-type semiconductor (Fig. 5.18a). As V_g increases, $|S|$ monotonically decreases from 490 μV K^{-1} to 90 μV K^{-1}. The observed S values reflect a bulk-like energy derivation of the parabolic-shaped DOS of the conduction band near the E_F. The observable S is roughly expressed as $S_{obs} = (\sigma_{s2DEG} \cdot S_{2DEG} + \sigma_{sBulk} \cdot S_{Bulk})/(\sigma_{s2DEG} + \sigma_{sBulk})$, where σ_s is the sheet conductance of each layer. In the present case, S_{2DEG} dominates S_{obs} due to the relation of $\sigma_{s2DEG} \gg \sigma_{sBulk}$. Here I like to discuss the t_{eff} of 2DEG at AlGaN/GaN. Figure 5.18b shows the calculated S of bulk GaN as a function of the n_{3D} at 300 K. The calculated line completely reproduces these reported values [47–49]. Then, we obtained the n_{3D} values using the observed S. Figure 5.18c shows calculated $t_{eff} = n_{2D}/n_{3D}$ as a function of the V_g. t_{eff} dramatically decreases from 20 nm to ~2 nm with V_g when $V_g < -6$ V (sub-threshold region), whereas it is almost saturated at a constant value (~2 nm), which agrees well with the previously reported 2DEG thickness [50], when $V_g > -6$ V.

The PF of the AlGaN/GaN 2DEG was calculated using the observed S, n_{3D} obtained from Fig. 5.18b, and the observed μ_{FE} as shown in Fig. 5.19. A high PF of ~9 mW m^{-1} K^{-2} with n_{3D} ~2.5 × 10^{19} cm^{-3} is calculated in high-mobility 2DEG at the AlGaN/GaN interface at room temperature, which is an order magnitude greater than that of doped GaN bulk [48] and a two- to sixfold increase compared to those of the state-of-the-art practical thermoelectric materials (1.5–4 mW m^{-1} K^{-2}). The carrier mobility of AlGaN-GaN 2DEG at

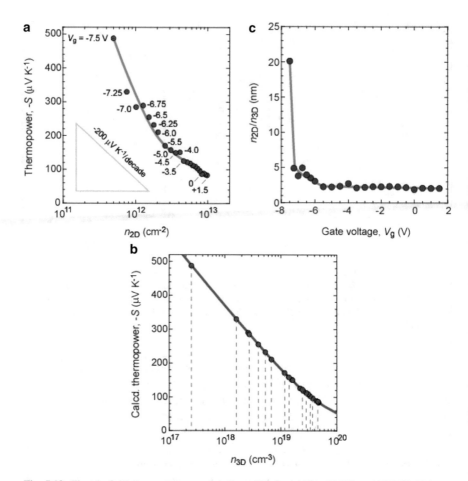

Fig. 5.18 Electric field thermopower modulation of high-mobility 2DEG at AlGaN/GaN interface. (**a**) Electric field modulated thermopower vs. sheet carrier concentration (n_s). (**b**) Calculated thermopower vs. volume carrier concentration (n_{3D}). (**c**) Effective thickness (n_{2D}/n_{3D}) as a function of the gate voltage (V_g). (Reprinted from [17] with permission from John Wiley and Sons)

$n_{3D} \sim 2.5 \times 10^{19}$ cm^{-3} is $\sim$1500 cm^2 V^{-1} s^{-1}, which is an order magnitude larger than that of conventional impurity doped bulk GaN ($\sim$125 cm^2 V^{-1} s^{-1}).

Finally, I would like to discuss the t_{eff} of 2DEG at AlGaN/GaN. The λ_D of GaN is $\sim$10 nm, which can be calculated by the following equation:

$$\lambda_D = \frac{h}{\sqrt{3 \cdot m* \cdot k_B \cdot T}}.$$

In our result, t_{eff} strides over λ_D. Thus, an enhanced S can be expected because it is theoretically predicted that a quantum well narrower than λ_D will exhibit

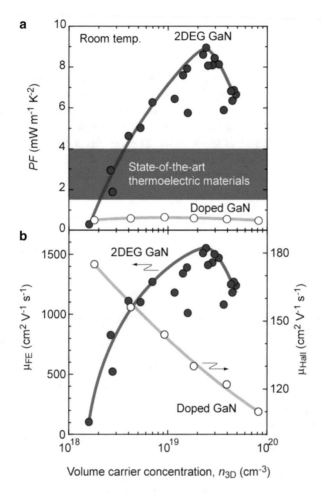

Fig. 5.19 High thermoelectric power factor of high-mobility 2DEG at AlGaN/GaN interface. Carrier concentration dependence of (**a**) PF and (**b**) carrier mobility (μ) for 2DEG GaN and conventional impurity doped n-type GaN. Maximized PF of the 2DEG with $n_{3D} \sim 2.5 \times 10^{19}$ cm^{-3} is ~ 9 mW m^{-1} K^{-2} at room temperature, which is an order magnitude greater than that of doped GaN bulk and a factor of 2–6 greater than that the state-of-the-art practical thermoelectric materials (1.5–4 mW m^{-1} K^{-2}). (Reprinted from [17] with permission from John Wiley and Sons)

an enhanced S [10, 11]. In the case of SrTiO$_3$-based 2DEG, a V-shaped upturn of S is observed when the 2DEG thickness is narrower than ~ 2 nm, clearly demonstrating the theory. However, such an S behavior is not observed in the present AlGaN/GaN 2DEG, indicating that any special effect of 2DEG does not contribute to the observed S. Thus, further study is required to clarify whether 2DEG is really effective to improve the thermoelectric performance or not.

5.7 Summary

In this manuscript, I have reviewed the points of our electric field thermopower modulation method of 2DEG using $SrTiO_3$, $BaSnO_3$, and AlGaN/GaN thin-film transistor structures. By combining volume carrier concentration dependence of thermopower, the electric field thermopower modulation results give the effective thickness of two-dimensional electron gas. This method is very useful to characterize the electronic structure of semiconductors especially the dimensionality. I hope this manuscript is useful to the materials scientists.

References

1. F.J. DiSalvo, Thermoelectric cooling and power generation. Science **285**(5428), 703–706 (1999)
2. G.J. Snyder, E.S. Toberer, Complex thermoelectric materials. Nat. Mater. **7**(2), 105–114 (2008)
3. T.J. Seebeck, Abh. K. Akad. Wiss. **265**, 1822 (1823)
4. K.F. Hsu, S. Loo, F. Guo, W. Chen, J.S. Dyck, C. Uher, T. Hogan, E.K. Polychroniadis, M.G. Kanatzidis, Cubic AgPbmSbTe2+m: Bulk thermoelectric materials with high figure of merit. Science **303**(5659), 818–821 (2004)
5. K. Biswas, J.Q. He, I.D. Blum, C.I. Wu, T.P. Hogan, D.N. Seidman, V.P. Dravid, M.G. Kanatzidis, High-performance bulk thermoelectrics with all-scale hierarchical architectures. Nature **489**(7416), 414–418 (2012)
6. L.D. Zhao, S.H. Lo, Y.S. Zhang, H. Sun, G.J. Tan, C. Uher, C. Wolverton, V.P. Dravid, M.G. Kanatzidis, Ultralow thermal conductivity and high thermoelectric figure of merit in SnSe crystals. Nature **508**(7496), 373 (2014)
7. L.D. Zhao, G.J. Tan, S.Q. Hao, J.Q. He, Y.L. Pei, H. Chi, H. Wang, S.K. Gong, H.B. Xu, V.P. Dravid, C. Uher, G.J. Snyder, C. Wolverton, M.G. Kanatzidis, Ultrahigh power factor and thermoelectric performance in hole-doped single-crystal SnSe. Science **351**(6269), 141–144 (2016)
8. S. Il Kim, K.H. Lee, H.A. Mun, H.S. Kim, S.W. Hwang, J.W. Roh, D.J. Yang, W.H. Shin, X.S. Li, Y.H. Lee, G.J. Snyder, S.W. Kim, Dense dislocation arrays embedded in grain boundaries for high-performance bulk thermoelectrics. Science **348**(6230), 109–114 (2015)
9. L.D. Hicks, T.C. Harman, X. Sun, M.S. Dresselhaus, Experimental study of the effect of quantum-well structures on the thermoelectric figure of merit. Phys. Rev. B **53**(16), 10493–10496 (1996)
10. L.D. Hicks, M.S. Dresselhaus, Effect of quantum-well structures on the thermoelectric figure of merit. Phys. Rev. B **47**(19), 12727–12731 (1993)
11. N.T. Hung, E.H. Hasdeo, A.R.T. Nugraha, M.S. Dresselhaus, R. Saito, Quantum effects in the thermoelectric power factor of low-dimensional semiconductors. Phys. Rev. Lett. **117**(3), 036602 (2016)
12. M.S. Dresselhaus, G. Chen, M.Y. Tang, R.G. Yang, H. Lee, D.Z. Wang, Z.F. Ren, J.P. Fleurial, P. Gogna, New directions for low-dimensional thermoelectric materials. Adv. Mater. **19**(8), 1043–1053 (2007)
13. Y.Q. Zhang, B. Feng, H. Hayashi, C.P. Chang, Y.M. Sheu, I. Tanaka, Y. Ikuhara, H. Ohta, Double thermoelectric power factor of a 2D electron system. Nat. Commun. **9**, 2224 (2018)
14. H. Ohta, S. Kim, Y. Mune, T. Mizoguchi, K. Nomura, S. Ohta, T. Nomura, Y. Nakanishi, Y. Ikuhara, M. Hirano, H. Hosono, K. Koumoto, Giant thermoelectric Seebeck coefficient of two-dimensional electron gas in SrTiO3. Nat. Mater. **6**(2), 129–134 (2007)

15. M. Cutler, N.F. Mott, Observation of Anderson localization in an electron gas. Phys. Rev. **181**(3), 1336 (1969)
16. K.P. Pernstich, B. Rossner, B. Batlogg, Field-effect-modulated Seebeck coefficient in organic semiconductors. Nat. Mater. **7**(4), 321–325 (2008)
17. H. Ohta, S.W. Kim, S. Kaneki, A. Yamamoto, T. Hashizume, High thermoelectric power factor of high-mobility 2D electron gas. Adv. Sci. **5**(1), 1700696 (2018)
18. A.V. Sanchela, T. Onozato, B. Feng, Y. Ikuhara, H. Ohta, Thermopower modulation clarification of the intrinsic effective mass in transparent oxide semiconductor BaSnO$_3$. Phys. Rev. Mater. **1**, 034603 (2017)
19. H. Ohta, Electric-field thermopower modulation in SrTiO3-based field-effect transistors. J. Mater. Sci. **48**(7), 2797–2805 (2013)
20. H. Ohta, T. Mizuno, S.J. Zheng, T. Kato, Y. Ikuhara, K. Abe, H. Kumomi, K. Nomura, H. Hosono, Unusually large enhancement of thermopower in an electric field induced two-dimensional electron gas. Adv. Mater. **24**(6), 740–744 (2012)
21. H. Ohta, Y. Sato, T. Kato, S. Kim, K. Nomura, Y. Ikuhara, H. Hosono, Field-induced water electrolysis switches an oxide semiconductor from an insulator to a metal. Nat. Commun. **1**, 118 (2010)
22. H. Ohta, Y. Masuoka, R. Asahi, T. Kato, Y. Ikuhara, K. Nomura, H. Hosono, Field-modulated thermopower in SrTiO3-based field-effect transistors with amorphous 12CaO center dot 7Al(2)O(3) glass gate insulator. Appl. Phys. Lett. **95**(11), 113505 (2009)
23. T. Okuda, K. Nakanishi, S. Miyasaka, Y. Tokura, Large thermoelectric response of metallic perovskites: Sr1-xLaxTiO3 (0 <= x <= 0.1). Phys. Rev. B **63**(11), 113104 (2001)
24. H. Ohta, K. Sugiura, K. Koumoto, Recent progress in oxide thermoelectric materials: p-type Ca3CO4O9 and n-type SrTiO3. Inorg. Chem. **47**(19), 8429–8436 (2008)
25. H. Ohta, Thermoelectrics based on strontium titanate. Mater. Today **10**(10), 44–49 (2007)
26. S. Ohta, T. Nomura, H. Ohta, M. Hirano, H. Hosono, K. Koumoto, Large thermoelectric performance of heavily Nb-doped SrTiO3 epitaxial film at high temperature. Appl. Phys. Lett. **87**(9), 092108 (2005)
27. S. Ohta, T. Nomura, H. Ohta, K. Koumoto, High-temperature carrier transport and thermoelectric properties of heavily La- or Nb-doped SrTiO3 single crystals. J. Appl. Phys. **97**(3), 034106 (2005)
28. M. Kawasaki, K. Takahashi, T. Maeda, R. Tsuchiya, M. Shinohara, O. Ishiyama, T. Yonezawa, M. Yoshimoto, H. Koinuma, Atomic control of the srtio3 crystal-surface. Science **266**(5190), 1540–1542 (1994)
29. H. Hosono, Y. Abe, An oxygen-effervescent aluminate glass. J. Am. Ceram. Soc. **70**(3), C38–C39 (1987)
30. H. Hosono, D.C. Paine, *Handbook of Transparent Conductors* (Springer, Berlin, 2011)
31. H.J. Kim, U. Kim, H.M. Kim, T.H. Kim, H.S. Mun, B.G. Jeon, K.T. Hong, W.J. Lee, C. Ju, K.H. Kim, K. Char, High mobility in a stable transparent perovskite oxide. Appl. Phys. Express **5**(6), 061102 (2012)
32. H.J. Kim, U. Kim, T.H. Kim, J. Kim, H.M. Kim, B.G. Jeon, W.J. Lee, H.S. Mun, K.T. Hong, J. Yu, K. Char, K.H. Kim, Physical properties of transparent perovskite oxides (Ba, La)SnO$_3$ with high electrical mobility at room temperature. Phys. Rev. B **86**(16), 165205 (2012)
33. P.V. Wadekar, J. Alaria, M. O'Sullivan, N.L.O. Flack, T.D. Manning, L.J. Phillips, K. Durose, O. Lozano, S. Lucas, J.B. Claridge, M.J. Rosseinsky, Improved electrical mobility in highly epitaxial La:BaSnO$_3$ films on SmScO$_3$ (110) substrates. Appl. Phys. Lett. **105**(5), 052104 (2014)
34. C. Park, U. Kim, C.J. Ju, J.S. Park, Y.M. Kim, K. Char, High mobility field effect transistor based on BaSnO3 with Al2O3 gate oxide. Appl. Phys. Lett. **105**(20), 203503 (2014)
35. U. Kim, C. Park, T. Ha, Y.M. Kim, N. Kim, C. Ju, J. Park, J. Yu, J.H. Kim, K. Char, All-perovskite transparent high mobility field effect using epitaxial BaSnO3 and LaInO3. Apl Mater. **3**(3), 036101 (2015)
36. S. Raghavan, T. Schumann, H. Kim, J.Y. Zhang, T.A. Cain, S. Stemmer, High-mobility BaSnO3 grown by oxide molecular beam epitaxy. Apl Mater. **4**(1), 016106 (2016)

37. S.J. Allen, S. Raghavan, T. Schumann, K.M. Law, S. Stemmer, Conduction band edge effective mass of La-doped BaSnO3. Appl. Phys. Lett. **108**(25), 252107 (2016)
38. J. Shin, Y.M. Kim, Y. Kim, C. Park, K. Char, High mobility BaSnO3 films and field effect transistors on non-perovskite MgO substrate. Appl. Phys. Lett. **109**(26), 262102 (2016)
39. C.A. Niedermeier, S. Rhode, S. Fearn, K. Ide, M.A. Moram, H. Hiramatsu, H. Hosono, T. Kamiya, Solid phase epitaxial growth of high mobility La: BaSnO3 thin films co-doped with interstitial hydrogen. Appl. Phys. Lett. **108**(17), 172101 (2016)
40. E. Moreira, J.M. Henriques, D.L. Azevedo, E.W.S. Caetano, V.N. Freire, U.L. Fulco, E.L. Albuquerque, Structural and optoelectronic properties, and infrared spectrum of cubic BaSnO3 from first principles calculations. J. Appl. Phys. **112**(4), 043703 (2012)
41. H.R. Liu, J.H. Yang, H.J. Xiang, X.G. Gong, S.H. Wei, Origin of the superior conductivity of perovskite Ba(Sr)SnO3. Appl. Phys. Lett. **102**(11), 112109 (2013)
42. B. Hadjarab, A. Bouguelia, M. Trari, Optical and transport properties of lanthanum-doped stannate BaSnO3. J. Phys. D. Appl. Phys. **40**(19), 5833–5839 (2007)
43. H.J. Kim, J. Kim, T.H. Kim, W.J. Lee, B.G. Jeon, J.Y. Park, W.S. Choi, D.W. Jeong, S.H. Lee, J. Yu, T.W. Noh, K.H. Kim, Indications of strong neutral impurity scattering in Ba(Sn,Sb)O-3 single crystals. Phys. Rev. B **88**(12), 125204 (2013)
44. D. Seo, K. Yu, Y.J. Chang, E. Sohn, K.H. Kim, E.J. Choi, Infrared-optical spectroscopy of transparent conducting perovskite (La,Ba)SnO3 thin films. Appl. Phys. Lett. **104**(2), 022102 (2014)
45. B.C. Luo, X.S. Cao, K.X. Jin, C.L. Chen, Determination of the effective mass and nanoscale electrical transport in La-doped BaSnO3 thin films. Curr. Appl. Phys. **16**(1), 20–23 (2016)
46. Y. Hori, Z. Yatabe, T. Hashizume, Characterization of interface states in Al2O3/AlGaN/GaN structures for improved performance of high-electron-mobility transistors. J. Appl. Phys. **114**(24), 244503 (2013)
47. M.S. Brandt, P. Herbst, H. Angerer, O. Ambacher, M. Stutzmann, Thermopower investigation of n- and p-type GaN. Phys. Rev. B **58**(12), 7786–7791 (1998)
48. A. Sztein, H. Ohta, J.E. Bowers, S.P. DenBaars, S. Nakamura, High temperature thermoelectric properties of optimized InGaN. J. Appl. Phys. **110**(12), 123709 (2011)
49. K. Nagase, S. Takado, K. Nakahara, Thermoelectric enhancement in the two-dimensional electron gas of AlGaN/GaN heterostructures. Phys. Status Solidi A **213**(4), 1088–1092 (2016)
50. Y.C. Kong, Y.D. Zheng, C.H. Zhou, Y.Z. Deng, S.L. Gu, B. Shen, R. Zhang, R.L. Jiang, Y. Shi, P. Han, Study of two-dimensional electron gas in AlN/GaN heterostructure by a self-consistent method. Phys. Status Solidi B Basic Res. **241**(4), 840–844 (2004)

Chapter 6
Transition-Metal-Nitride-Based Thin Films as Novel Thermoelectric Materials

Per Eklund, Sit Kerdsongpanya, and Björn Alling

6.1 Introduction

The early transition-metal and rare-earth nitrides, primarily based on ScN and CrN, have emerged as an unexpected class of materials for energy harvesting by thermoelectricity and piezoelectricity and more generally for conversion of heat or mechanical energy to electricity. Largely ignored for these purposes until around 2010, this class of materials has seen a number of fundamental advances, among those the discoveries of exceptionally high piezoelectric coupling coefficient in (Sc,Al)N alloys [1] and of remarkably high thermoelectric power factors of ScN-based [2–6] and CrN-based [7–10] thin films. These materials also constitute well-defined model systems for investigating thermodynamics of mixing for alloying and nanostructural design for optimization of phase stability and band structure. These

This chapter is an adapted and updated version of the review article "Transition-metal-nitride-based thin films as novel energy harvesting materials", P. Eklund, S. Kerdsongpanya, B. Alling J. Mater. Chem. C 4, 3905–3914 [used here under CC BY 3.0 license]. When referencing the present work, please cite also the original article.

P. Eklund (✉)
Department of Physics, Chemistry, and Biology (IFM), Linköping University, Linköping, Sweden
e-mail: per.eklund@liu.se

S. Kerdsongpanya
Department of Physics, Chemistry, and Biology (IFM), Linköping University, Linköping, Sweden

Western Digital Corporation, Magnetic Heads Operations, Materials Science Labs, Ayutthaya, Thailand

B. Alling
Department of Physics, Chemistry, and Biology (IFM), Linköping University, Linköping, Sweden

Max-Planck-Institut für Eisenforschung GmbH, Düsseldorf, Germany

© Springer Nature Switzerland AG 2019
P. Mele et al. (eds.), *Thermoelectric Thin Films*,
https://doi.org/10.1007/978-3-030-20043-5_6

121

features have implications for and can be used for tailoring of thermoelectric and piezoelectric properties.

The process of energy harvesting is the capture of energy from ambient sources and storage and/or application for use as power sources. There is a wide range of ambient sources, such as solar, wind, electromagnetic radiation, mechanical (kinetic) energy, and thermal energy. Energy harvesting differs conceptually from, e.g., oil and coal power, fuel cells, or batteries that involve active combustion of a fuel or conversion of stored chemical energy to electricity. Furthermore, the term energy harvesting is typically reserved for capturing energy for powering small, low-power devices, usually off-grid or otherwise autonomous. The term does not include, e.g., solar-power and wind-power plants, although the fundamental concept is the same.

In this chapter, we review the ScN- and CrN-based transition-metal nitrides for thermoelectrics (harvesting of ambient heat), drawing parallels with piezoelectricity (harvesting of mechanical vibrations). It is further intended as an example of general strategies for tailoring of thermoelectric properties by integrated theoretical-experimental approaches. This chapter is an adapted and updated version of the review article "Transition-metal-nitride-based thin films as novel energy harvesting materials" [used here under CC-BY 3.0 unported license] [11]. When referencing the present work, please cite also the original article.

6.2 Brief Introduction to Thermoelectricity

Thermoelectric devices harvest thermal energy (temperature gradients) into electricity and can also be used for environment-friendly refrigeration, without moving parts or malign liquids or gases [12]. Most other conversion systems (such as power plants) become less efficient as they are scaled down in size and power, but thermoelectrics benefit from low- to medium-power and -size application. Thus, potential contributions of thermoelectrics are in applications with relatively low power levels used in large numbers (for example, in personal computers, automotive applications, and consumer electronics). Furthermore, an important trend is the use of thermoelectric devices as wearable and/or mechanically flexible power sources [13] with much focus on organic thermoelectrics [14–16] but also a number of recent studies on carbon-nanotube-based materials [17, 18], and on mechanically flexible but fully inorganic ceramic materials and devices [19–21].

The efficiency of a thermoelectric material at a temperature T is related to the dimensionless figure of merit ZT, where $Z = S^2\sigma/\kappa$. Here, σ and κ are the electrical and thermal conductivities, respectively, and S is the Seebeck coefficient $\Delta V/\Delta T$, i.e., the voltage in response to a temperature gradient. The thermal conductivity is given by $\kappa = \kappa_l + \kappa_e$, where the subscripts l and e denote the lattice (phonon) and electron contributions, respectively. Thus, a high ZT requires a good electrical conductor with high Seebeck coefficient but low thermal conductivity [22]. For traditional thermoelectric materials like tellurides and antimonides, $ZT \approx 1$ at room temperature. At first glance, it seems easy to increase ZT by, e.g., increasing the conductivity by a factor of 2–4. However, basic transport theory in solids implies that S, σ, and κ are interrelated; increasing σ by increasing the charge carrier concentration results in lower S and increased κ, yielding no improvement in ZT.

The delicate interdependence of the three parameters S, σ, and κ requires novel approaches to advance the field of thermoelectrics, which has led to extensive efforts on nanostructural design [23, 24]. Predictions in the mid-1990s suggested that ZT could be enhanced by quantum confinement [25]. In 2001, thermoelectric superlattice devices of Bi_2Te_3/Sb_2Te_3 with remarkably high ZT (~2.4) caused a surge of interest [26]. Materials with such high ZT, however, are restricted to laboratory devices, which has prevented their success in practice [27]. At present, the main achievement of nanostructuring is reduction of the lattice thermal conductivity [28, 29] rather than improvements due to quantum confinement of charge carriers [23, 24].

Therefore, there is a need to introduce mechanisms that, in addition to reducing κ, also enhance the thermoelectric power factor ($S^2\sigma$). One approach is to consider theoretically what band structure a hypothetical material should have to maximize ZT. Mahan and Sofo [30] predicted this in the 1990s, and others have more recently refined the picture [31]. For a given κ_l, the ideal transport-distribution function that maximizes ZT is a bounded delta function, approximately realized in practice as a sharp function with a large slope in the density of states (DOS) at the Fermi level E_F [24, 25]. These approaches are the base for modern strategies for the development of thermoelectrics: band structure optimization to emulate the ideal band structure, combined with nanostructural design to reduce the thermal conductivity.

6.3 The Early Transition-Metal Nitrides

6.3.1 Overall Trends

The early transition-metal nitrides—and their alloys—based on group-4 (Ti, Zr, and Hf) or group-5 (V, Nb, and Ta) metals, are long-established in applications as hard, wear-resistant coatings, with TiN being the archetype [32–34]. While they are hard and exhibit other typical ceramic properties, these nitrides are metallic in nature with respect to electrical properties and in fact very good conductors, with typical resistivities in the approximate range 10–30 $\mu\Omega$cm (in comparison, noble metals have resistivities of a few $\mu\Omega$cm, and Ti metal about 40 $\mu\Omega$cm) [35]. Thus, they find extensive use as conducting permanent contact layers and diffusion barriers in microelectronics.

Reducing the valency by one from Ti or increasing it by one from V, i.e., moving to groups 3 or 6 in the periodic table, results in drastically altered properties. The cubic rock-salt-structured ScN and CrN are both narrow-bandgap semiconductors. Sc has three valence electrons that together with the three $2p$ valence electrons of N complete the filling of the bonding states formed by nearest neighbor hybridization of mainly N $2p$ and Sc $3d$ e_g with some Sc $4s$ character. The $3d$ t_{2g}-orbitals, on the other hand, are completely empty, in contrast to the group-4 and -5 transition metals, and the Fermi level drops below the conduction band edge, causing ScN to become semiconducting. In CrN, with three more electrons than ScN, the nonbonding Cr $3d$ t_{2g}-band is half-filled. This causes a spin splitting of the band which gives Cr atoms of CrN a distinct local magnetic moment approaching 3 μ_B [36]. As a consequence, also in this material, the Fermi level falls into a bandgap, this time between occupied spin-up nonbonding Cr $3d$ t_{2g} and an unoccupied mixture of mostly antibonding spin-up Cr $3d$ e_g, and nonbonding spin-down Cr $3d$ t_{2g} [37, 38].

The effect on resistivity is illustrated in Fig. 6.1, from an early study by Gall et al. [39] who investigated $Ti_xSc_{1-x}N$ epitaxial thin films. Pure TiN is a good conductor and exhibits a typical room temperature resistivity value around 20 $\mu\Omega$cm, and the archetypical metallic temperature dependence with constant resistivity at low temperature dominated by scattering from vacancies, defects, and impurities. As temperature increases, the resistivity increases linearly with the scattering dominated by electron-phonon coupling. For low Sc content in the $Ti_xSc_{1-x}N$ alloy, this behavior is initially retained, but for higher Sc content, an increase in resistivity is observed at cryogenic temperature and for pure ScN this effect is dominant indicating semiconducting behavior. It needs to be stressed, though, that this is highly dependent on impurities and dopants.

Fig. 6.1 Electrical properties of ScN, TiN, and ScTiN alloys. Adapted from Gall et al. [39] (Copyright American Institute of Physics, used with permission)

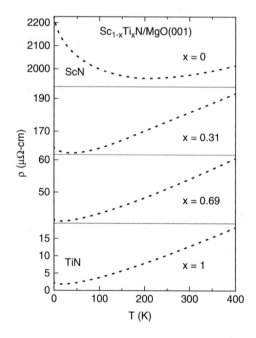

6.3.2 ScN

ScN, like the group-4 and group-5 transition-metal nitrides, is an interstitial (cubic NaCl structure) nitride following Hägg's rule; that interstitial nitrides and carbides are formed when the radii ratio between the nonmetal and metal atoms is smaller than 0.59 [40]. ScN has similar hardness as the other transition-metal nitrides, 21 GPa, and is stable at high temperature (melting point 2550 °C), while it is prone to oxidation if used in air above 600 °C [41–43].

As stated above, ScN is a narrow-bandgap semiconductor. The fundamental bandgap is ~0.9 eV and the direct (optical bandgap) is ~2.1 eV. This was, however, a topic of debate for a rather long time with numerous studies yielding conflicting results as to whether ScN was a semiconductor, a semimetal, or even a metal [44–47]. The reason for these discrepancies is that it is challenging to produce pure ScN. Sc has a high affinity to oxygen [48] and, if not synthesized in a pure ultrahigh-vacuum environment, can readily contain large amounts of oxygen impurities, as well as contaminations from residual hydrocarbons. Free carriers from impurities can result in large inaccuracies in determinations of optical bandgaps [47]. Furthermore, processing of scandium ore involves a purification step with fluoride reduction [49], which results in scandium raw materials often containing fluorine impurities.

There are relatively many studies on thin-film growth of ScN. Among the methods used, magnetron sputter deposition [2–6, 39, 47, 50–52], chemical vapor deposition [41, 53], and molecular beam epitaxy [54–59] are the most common. Irrespective of method, the aspects of reactivity and oxygen and/or fluorine uptake (or other impurities) are essential in thin-film growth of ScN, stressing the need for a pure environment.

For a transition-metal nitride, ScN exhibits an anomalously high thermoelectric power factor [2, 3] $S^2\sigma$. $S^2\sigma$ is in the range 2.5–3.3 $Wm^{-1} K^{-2}$, well on par with established thermoelectric materials such as PbTe [60]. This is illustrated in Fig. 6.2, where ScN (our data from Ref. [2] and the results of Burmistrova et al. [3]) is shown in relation to a typical value for n-type PbTe. In comparison, the power factor of Bi_2Te_3 is somewhat higher at above 4 $Wm^{-1} K^{-2}$. The thermal conductivity of ScN, though, is much higher than for these tellurides, in the range 8–12 $Wm^{-1} K^{-1}$ [3, 5, 41, 61], and would need to be drastically reduced to enable application of ScN as a thermoelectric material; strategies for addressing this are discussed in Sect. 6.5 below. These tellurides are benchmark thermoelectric materials. Nonetheless, the scarcity [62] of Te as well as legislative restrictions on the use of Pb limits their applicability outside niche applications. Hence, much effort is devoted to developing alternative materials. The early transition-metal nitrides are a class that was not much considered for this purpose until just a few years ago. From an application point of view, CrN-based materials are closer to application than ScN-based ones, since the former are abundant, relatively inexpensive, and can readily be made in large quantities by standard processing techniques in both thin films and bulk.

The unexpectedly high thermoelectric power factor of ScN can be explained based on band structure features caused by impurities. A conceptual illustrative example is shown in Fig. 6.3 (adapted from Ref. [63]), where first principles calculations show that the combination of (in this example) C dopants and N vacancies in ScN introduces a sharp variation in the density of states at the Fermi level. As described in Sect. 6.2, to maximize ZT, the transport-distribution function should be a bounded delta function (for a given phonon κ), realized in practice as a large slope in the density of states near the Fermi level. Thus, the electronic structure of ScN—including vacancies and impurities in the level of ~1 at. %—can mimic the ideal theoretical transport-distribution function, yielding a high power factor [63, 64]. The same conclusions are drawn from calculations with O or F dopants [3, 63].

6.3.3   CrN

CrN is well known from applications as hard coatings, given its high hardness of 28 GPa and good resistance to wear and corrosion [65–75]. It can also be readily synthesized in bulk [76, 77]. In addition to these properties, CrN exhibits a magnetic phase transition [36, 78, 79]. Above the Néel temperature (T_N) of 286 K, CrN is paramagnetic with cubic (NaCl) crystal structure with a lattice parameter reported to range from 4.135 to 4.185 Å [80–83]. Below this temperature, the structure is antiferromagnetic and orthorhombic.

The electrical properties of CrN can vary greatly; for example, various studies show resistivity values ranging from 1.7 to 350 mΩcm [7, 72, 77, 84]. As for

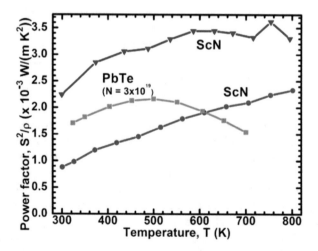

Fig. 6.2 Thermoelectric power factor $S^2\sigma$ of ScN (the bottom curve shows our first data from Ref. [2] and the top curve are the data of Burmistrova et al. [3]. The example data for PbTe are adapted from Sootsman et al. [60] (From Ref. [11], used under CC-BY license)

Fig. 6.3 Example of effects
of vacancies and dopants on
the band structure of ScN
(adapted from Ref. [63]).
Bottom: pure, stoichiometric
ScN (note the inaccurate
bandgap determination with
GGA. Second from bottom:
1% N vacancies. Second from
top: 1% C dopants. Top: Both
C dopants and N vacancies.
(From Ref. [11], used under
CC-BY license)

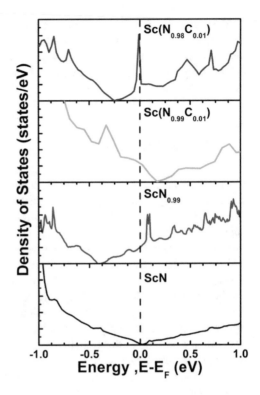

the reported temperature-dependent behavior of the electrical resistivity around the
cubic-to-orthorhombic phase transition [76], there is typically a jump in resistivity
between these two semiconducting phases, though there are reports of metallic
behavior for the orthorhombic phase [80]. Part of this apparent discrepancy can
be attributed to CrN being a narrow-bandgap semiconductor, where the presence
of N vacancies may act as effective dopants yielding high electron concentrations
and metallic-like behavior below the Néel temperature [37]. In epitaxial thin films,
it is possible to stabilize the cubic phase and suppress the phase transition to the
orthorhombic phase [85] [86] [82]. In addition, Gall et al. [86] suggested that the
conducting behavior in CrN films is a hopping conduction mechanism, and the band
gap of CrN depends on the correlation energy.

For thermoelectric properties, CrN exhibits high Seebeck coefficients of typically
around 135 mV/K around room temperature and up to 200 mV/K at 600 K [77].
Also, its thermal conductivity is moderate at ~1.7 $Wm^{-1} K^{-1}$ (~1/5 of that of
ScN). Nonetheless, the electrical resistivity is relatively high in pure form, because
the localized $3d$ orbitals of Cr give large effective masses causing high Seebeck
coefficients and resistivities. This was addressed by Quintela et al. [8] who annealed
as-deposited films in ammonia gas for 2 h at 800 °C to ensure that the films
were fully stoichiometric and to improve the crystalline quality, yielding a large
improvement in Seebeck coefficient and a hundredfold reduction in resistivity.

This further underscores that the early transition-metal nitrides hold unexpected promise as novel thermoelectric materials. However, in their pure form, neither ScN nor CrN are likely to reach all the way; experimental strategies and theoretically guided design approaches for reduction of the thermal conductivity with retained or increased power factors are needed.

6.4   Theoretical Methodology

Electronic structure calculations based on density functional theory (DFT) [87, 88] and standard approximations for the exchange-correlation energies like the local density approximation (LDA) and the different flavors of the generalized gradient approximation (GGA) [89] are very efficient and most often accurate tools for phase stability and ground state-related properties for most classes of materials. However, both the existing cases where this framework is insufficient in treatment of the quantum electronic problem, and perhaps more frequently, the neglect in theoretical studies of relevant vibrational, magnetic, and structural disorder, present in reality, is an obstacle in first principles-based calculations for real-world materials science problems. In fact, the early $3d$ transition-metal nitrides constitute an interesting illustration of several of these challenges. In ScN, calculations of the bandgap using the Kohn-Sham orbital gap in GGA gives basically a zero-gap semiconductor, an underestimation effect well known and found for practically all semiconductors. This is seen in the bottom curve in Fig. 6.2, from Ref. [63]. In VN, the rock-salt structure observed at room temperature and above is actually dynamically stabilized by anharmonic lattice vibrations and unstable at low temperature [90]. This can lead to unphysical large atomic relaxation effects in VN-based systems if static, 0 K, calculations are performed without symmetry constraints, e.g., with point defects or alloys. Finally, in CrN, the effect of strong electron correlation is treatable on the level of LDA + U [37, 91, 92]. However, most importantly the magnetic degree of freedom has caused considerable debate. In particular, the bulk modulus of paramagnetic rock-salt phase of CrN has been modeled using non-spin polarized calculations [93]. However, the existence of local finite Cr moments, also above the magnetic ordering temperature, is crucial to consider in the theoretical modeling [94], otherwise the bulk modulus is greatly overestimated. This model, including local Cr moments in the paramagnetic phase, was later independently confirmed by new sets of experiments [95]. The simultaneous presence of lattice vibrations and disordered magnetic moments in paramagnetic rock-salt CrN causes further challenges for quantitative modeling. In fact, due to this complexity, the material has become a benchmark case for theoretical method development in the field [96–99]. A recent example is the theoretical discovery that dynamical spin-lattice coupling is the explanation for relatively low thermal conductivity in the paramagnetic state of CrN [100]. A similar finding of the importance of magnetism was also made for the bulk modulus of Cr_2AlC and other Cr-containing so-called MAX phases [101, 102]. The crucial importance of the details of the nitrogen content and oxygen contamination levels pointed out in the previous section also causes concern for the theoretical modeling of CrN as it couples with the vibrational and magnetic degrees of freedom [103].

The current outstanding issue in this line of theoretical development is associated with the unclear, material-specific, timescale for the propagation of the magnetic state as compared to the dynamics of the lattice. Recently, several methodological obstacles have been overcome, and methods for constrained local moment's calculations [104] and the derivation of Heisenberg-type exchange interactions [105], within the supercell-based plane-wave electronic structure frameworks needed for ab initio molecular dynamics now exist. The stage is set for a direct combination of molecular and spin dynamics in an effective ab initio manner with minimum or no free parameters. It would open up for first principles-based calculations of lattice and magnetic thermal conductivity in the paramagnetic phase of magnetic materials, such as CrN.

In the case of first principles modeling of substitutionally disordered nitride alloys, e.g., $Cr_{1-x}Al_xN$ and $Ti_{1-x}Al_xN$ [106], the configurational problem arises as the crystallographic unit cell is no longer sufficient to describe the material. In a completely random alloy, the components are stochastically distributed on the lattice sites, metal sublattice in the case of a nitride alloy, implying lack of long range order and existence of many different local chemical environments of the atoms. The most reliable method to model such materials is the Special Quasirandom Structure (SQS) method [107] introduced for transition-metal nitride alloys in a study of $Ti_{1-x}Al_xN$ [108]. Using the SQS approach the mixing thermodynamics of the alloys can be directly modeled within a mean-field approximation for the configurational entropy. Also, e.g., the piezoelectric properties can be calculated directly [109, 110]. For the alloys of transition-metal nitrides, and group-13 nitrides like AlN, such modeling has revealed important information about the mixing trends, i.e., if the supersaturated alloys obtained in the out-of-equilibrium synthesis at low temperature, will phase separate, order, or stay as a solid solutions when subject to the temperatures needed to induce metal sublattice diffusion, e.g., in several (Sc,M)N [111] and (Cr,M)N [112] alloys.

It should be noted that real alloys under equilibrium conditions always display some degree of partial short-range ordering or short-range clustering. This can of course also be the case for a metastable supersaturated solid solution grown with out-of-equilibrium techniques. However, in lack of a priori knowledge of such tendencies, the ideal random SQS approach is a well-defined, unbiased starting point for, e.g., more intricate cluster-expansion approaches of the configurational thermodynamics [113]. Outstanding issues here include the difficulty to include vibrational free energy, and in particular anharmonic contributions, into the configurational thermodynamics analysis in an accurate and efficient manner [114].

With a reliable state-of-the art theoretical description of the materials equilibrium properties, the door opens for accurate calculations of properties. However, such calculations involve drastically different levels of complexity depending on which property that is needed. The properties needed for predicting the piezoelectric response of a material are second-order strain-derivatives of ground state energies, elastic constants, and first-order strain-derivatives of polarization [115]. These are relatively straightforward to calculate accurately from first principles with the complexities arising mostly from their tensorial nature, where care must be taken in the case of disordered alloys [116].

The properties needed for understanding thermoelectric behavior of a material, on the other hand, are quite challenging to derive directly and accurately from first principles, because the thermoelectric figure of merit includes both electronic and thermal transport and the entropy involves nonequilibrium transport processes. Ab initio calculation of thermoelectric parameters is addressed by Boltzmann transport theory [30, 117], but involves an unknown scattering parameter, the relaxation time τ. For the Seebeck coefficient (and Hall coefficient), τ cancels out if it is isotropic and constant with respect to energy. However, electrical and (electronic) thermal conductivities can only be determined either as a function of τ or by fitting to experimentally determined values [118] of τ, placing a substantial limitation on these computational approaches. Such calculations of the latter parameters are therefore not truly ab initio but restricted to materials for which experimental data of τ (or parameters from which τ can be calculated) are available. Ongoing method development is therefore devoted to finding methods for computing these from first principles. Example is the recent work of Faghaninia et al. [119] who derived an ab initio approach for computing these properties in the low-electric-field limit, and efforts to incorporate low-temperature effect of phonon drag [120, 121].

6.5   Ternary Systems

As discussed above, ScN and CrN are promising for thermoelectrics, but a reduced thermal conductivity of ScN or reduced electrical resistivity of CrN would be required for actual applications. This can be addressed in a $Cr_{1-x}Sc_xN$ solid solution, which is thermodynamically stable at high temperature in cubic NaCl-structured form [122]. The fact that the $3d$ orbitals in Sc are empty can be exploited as a means of delocalizing the electrons in $3d$ orbitals, resulting in electrical conductivity reduction [122] and possibly also thermal conductivity reduction due to alloy scattering. We have recently shown that the Seebeck coefficient of Sc-rich $Cr_{1-x}Sc_xN$ solid solution epitaxial thin films does indeed increase compared to pure ScN and that the thermoelectric properties of CrN are largely retained in Cr-rich $Cr_{1-x}Sc_xN$ solid solutions [122].

ScN-based solid solutions are further important, because of the interest caused by the exceptionally high piezoelectric coupling coefficient in (Sc,Al)N alloys [1, 123]. (Sc,Al)N and (Sc,Ga)N alloys were recently reviewed by Moram and Zhang [124] and the reader is referred there. (Sc,Y)N was investigated by Gregoire et al. [125] and (Sc,Mn)N by Saha et al. [126]. A particular important example here is (Sc, Mg)N, where the group-2 element Mg acts as acceptor and, if introduced in sufficient amount, can switch ScN to a p-type material [63, 127, 128]. This is also an illustrative example of the importance of the content of oxygen and other impurities in ScN. In the (Sc,Mg)N films of Saha et al. [127, 128], the low impurity content allowed isolating the effect of Mg doping, yielding a p-type material. In contrast as studied by Tureson et al. [129], in ScN films with higher O content with Mg introduced by ion implantation, the defects introduced showed a large reduction in thermal conductivity. However, with oxygen acting as dopant and cancelling out the acceptor role of Mg, the material remained n-type.

Alloy scattering is one of the standard strategies for thermoelectric materials for reduction of the lattice thermal conductivity; other approaches are superlattices, nanoinclusions, or grain boundaries [22, 23, 130–132]. Furthermore, the peaks in the density of states at the Fermi level causing high Seebeck coefficient is traditionally associated with reduced thermodynamic stability [133, 134]. For this reason, the search for optimal thermoelectric materials may be fruitful among metastable materials synthesized with far-from-equilibrium techniques, such as magnetron-sputtered metastable nitride thin film alloys, with the reservation that the metastable nature of such materials would place a limit on high-temperature long-term use.

ScN- and CrN-based systems are interesting model systems for these general research questions [111]. In particular, Sc is naturally isotope-pure, thus lacking isotope reduction of thermal conductivity. Consequently, the possibilities to substantially reduce the thermal conductivity by alloying or nanostructural engineering are particularly promising in this material. If the thermal conductivity can be reduced, ScN-based materials could potentially be applied at elevated temperatures, where bulk diffusion can be activated and the thermodynamics of mixing between ScN and the alloying or superlattice component becomes relevant. Superlattices might intermix, alloys could order or phase separate, and nanostructures might be dissolved in the matrix. All these processes will most likely affect thermoelectric properties.

Superlattices are of great interest for thermoelectrics, since they may allow for both the reduction of the lattice thermal conductivity and the quantum confinement of electrons. The first thermoelectric superlattice devices were made from combinations of the semiconductors Bi_2Te_3/Sb_2Te_3 [26]. A different approach is to combine the high electron concentrations of ultrathin metallic layers (e.g., TiN or ZrN) inserted between semiconductor barriers (e.g., CrN, ScN). The sharp asymmetry in the conduction electron distribution near the Fermi energy may be achieved for possible substantial improvements in ZT [135]. Furthermore, it has been demonstrated that the high interface density in a superlattice can reduce the thermal conductivity in ScN/(Zr,W)N superlattices [61]. This is illustrated in Fig. 6.4, from Rawat et al. [61].

A theoretically guided approach to implementing these strategies are to use density functional theory calculations to investigate the effect of mixing thermodynamics in order to determine phase stability of ScN-based solid solutions of relevance for lattice thermal conductivity reduction. Our results demonstrated [111] that at 800 °C the free energy of mixing for (Sc,Y)N, (Sc,La)N, (Sc,Gd)N, and (Sc,In)N exhibits a thermodynamic tendency for phase separation at high temperature. In addition, for the 50:50 Sc:M (M = V, Nb, or Ta) ratio, the (Sc,V)N, (Sc,Nb)N, and (Sc,Ta)N exhibit a stable ternary inherently nanolaminated phase

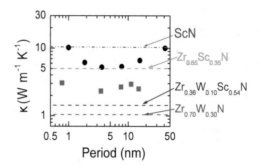

Fig. 6.4 Illustration of superlattice reduction in thermal conductivity, from Rawat et al. [61]. Cross-plane thermal conductivity of 300 nm thick ZrN/ScN (dots) and $Zr_{0.64}W_{0.36}N/ScN$ (squares) multilayers. Superimposed on the plot are horizontal lines corresponding to the experimentally determined lattice component of thermal conductivity, i.e., the alloy limit of different alloys of ZrN, ScN, and W_2N. (Copyright American Institute of Physics, used with permission)

[136] with the $ScTaN_2$-type structure. On the other hand, at 800 °C, the (Sc,Ti)N, (Sc,Zr)N, (Sc,Hf)N, and (Sc,Lu)N are thermodynamically stable in disordered B1 (NaCl) solid solutions, rather than in the ordered solid solutions which are stable at 0 K. This last point is shown in Fig. 6.5 (from Ref. [111]), which shows (Fig. 6.5a) a comparison of the calculated mixing enthalpies of substitutionally disordered solid solution, ordered solid solutions, and $ScTaN_2$-type structure phase of (Sc,M)N, as a function of MN content where M = Ti, Zr, and Hf, and (Fig. 6.5b) calculated equilibrium lattice parameter for the rock-salt (B1) solid solution as a function of MN content.

These results enabled us to suggest suitable materials for the different possible strategies for reduction of the lattice thermal conductivity of ScN. Since the heavy element Lu has a mixing tendency with ScN and has the same number of valence electrons, it is an appropriate choice for solid solution reduction of the thermal conductivity. YN, LaN, GdN, AlN, GaN, and InN have a thermodynamic tendency for phase separation with ScN and thus constitute good alloying elements if decomposition to form nanoinclusions is the strategy. The three former can be used for superlattices, since they, in addition, are isostructural with ScN. That, in combination with the thermodynamic tendency for phase separation, would render the superlattice structure stable. The wurtzite AlN, GaN, and InN are not suited for this purpose because of the difference in crystal structure, but their alloys with ScN or other NaCl-structure nitrides can be as described below [137–139]. The mixing thermodynamics of these alloy systems can be understood by considering the effect of the factors of volume mismatch, favoring phase separation, and an electronic structure effect of delocalization of extra d-electrons to empty Sc $3d$-t_{2g} states, favoring mixing.

Important experimental demonstrations of these principles are the works of Saha et al. who stabilized cubic (Sc,Al)N lattice-matched [138] to TiN and (Ti,W)N in TiN/(Sc,Al)N and (Ti,W)N/(Sc,Al)N superlattices exhibiting enhanced hardness

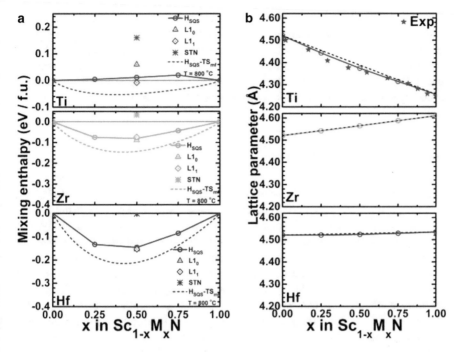

Fig. 6.5 (a) Comparison of the calculated mixing enthalpies of substitutionally disordered solid solution, ordered solid solutions, and ScTaN$_2$-type structure phase of Sc$_{1-x}$M$_x$N, as a function of MN content where M = Ti, Zr, and Hf, respectively. (b) Calculated equilibrium lattice parameter for rocksalt (B1) Sc$_{1-x}$M$_x$N solid solution as a function of MN content where M = Ti, Zr, and Hf, respectively. The black line indicates Vegard's rule. For (Sc,Ti)N, experimental data from Gall et al. [39] are shown with stars. From Ref. [111] (Copyright American Institute of Physics, used with permission)

[139], large reduction in thermal conductivity [140], and have been studied for plasmoic properties [141], diffusion mechanisms [142], and phase transformations [143]. Here, the combination of superlattice scattering and heavy element alloying (with W) allowed for thermal conductivities down to 1.7 Wm^{-1} K^{-1} (compared to 8–10 Wm^{-1} K^{-1} for pure ScN). Nonetheless, as discussed above, these superlattice structures are metastable and limited in use to the medium-temperature range. Around 800 °C, where bulk diffusion becomes dominant, their long-term thermal stability would be compromised due to intermixing. This was demonstrated by Schreoder et al. who showed that TiN/(Sc,Al)N superlattices intermix heavily at elevated temperature [137]. For a more detailed discussion, please refer to the recent review of Saha et al. [144].

Finally, we note that recently theoretical calculations have been used to suggest further alternatives to the ScN- and CrN-based semiconducting systems. In particular, by combining group-4 transition metals with a group-2 alkaline-earth metal in equal amounts, novel semiconducting systems have been predicted, like Ti$_{0.5}$Mg$_{0.5}$N [145] and the wurtzite-structure (TM$_{0.5}$,M$_{0.5}$)$_x$Al$_{1-x}$N alloys are investigated for piezoelectric properties [146, 147]. These recent studies demonstrate the superior speed with which computationally based approaches can scan large pools of complex, uncharted materials and suggest candidates for a given property, before experimental verification is pursued. This emphasizes the promise for a future of theoretically driven materials' discoveries, at least in the cases where the theoretical accuracy and methodological reliability are well established.

6.6 Concluding Remarks

We have reviewed the present state of research on early transition-metal nitrides, primarily based on ScN and CrN, for thermoelectric energy harvesting. These materials also constitute well-defined model systems for theory-guided approaches investigating thermodynamics of mixing for alloying and nanostructural design for optimization of phase stability, band structure, and thermal conductivity in order to improve thermoelectric properties. This is most notable as thermoelectric properties per se—unlike piezoelectric properties—are challenging to reliably calculate by ab initio methods. This can be used to guide the implementation of strategies for reduction of the lattice thermal conductivity; alloy scattering, superlattices, and nanoinclusions.

Acknowledgments The authors acknowledge funding from the European Research Council under the European Community's Seventh Framework Programme (FP/2007–2013)/ERC grant agreement no 335383, the Swedish Government Strategic Research Area in Materials Science on Functional Materials at Linköping University (Faculty Grant SFO-Mat-LiU No 2009 00971, the Knut and Alice Wallenberg Foundation under the Wallenberg Academy Fellows program, Swedish Foundation for Strategic Research (SSF) through the Future Research Leaders 5 (P. E.) and 6 (B. A.) program, the Swedish Research Council (VR) under projects no 2016-03365, 621-2012-4430, project no 621-2011-4417, and International Career Grant 330-2014-6336.

References

1. M. Akiyama, T. Kamohara, K. Kano, A. Teshigahara, Y. Takeuchi, N. Kawahara, Adv. Mater. **21**, 593 (2009)
2. S. Kerdsongpanya, N. Van Nong, N. Pryds, A. Zukauskaite, J. Jensen, J. Birch, J. Lu, L. Hultman, G. Wingqvist, P. Eklund, Appl. Phys. Lett. **99**, 232113 (2011)
3. P.V. Burmistrova, J. Maassen, T. Favaloro, B. Saha, S. Salamat, Y.R. Koh, M.S. Lundstrom, A. Shakouri, T.D. Sands, J. Appl. Phys. **113**, 153704 (2013)
4. P.V. Burmistrova, D.N. Zakharov, T. Favaloro, A. Mohammed, E.A. Stach, A. Shakouri, T.D. Sands, J. Mater. Res. **30**, 626 (2015)
5. S. Kerdsongpanya, O. Hellman, B. Sun, Y.K. Koh, J. Lu, N. Van Nong, S.I. Simak, B. Alling, P. Eklund, Phys. Rev. B **96**, 195417 (2017)
6. A. le Febvrier, N. Tureson, N. Stilkerich, G. Greczynski, P. Eklund, J. Phys. D. Appl. Phys. **52**(3), 035302 (2018). https://doi.org/10.1088/1361-6463/aaeb1b
7. C.X. Quintela, B. Rodriguez-González, F. Rivadulla, Appl. Phys. Lett. **104**, 022103 (2014)

8. C.X. Quintela, J.P. Podkaminer, M.N. Luckyanova, T.R. Paudel, E.L. Thies, D.A. Hillsberry, D.A. Tenne, E.Y. Tsymbal, G. Chen, C.-B. Eom, F. Rivadulla, Adv. Mater. **27**, 3032 (2015)
9. M.A. Gharavi, S. Kerdsongpanya, S. Schmidt, F. Eriksson, N.V. Nong, J. Lu, B. Balke, D. Fournier, L. Belliard, A. le Febvrier, C. Pallier, P. Eklund, J. Phys. D. Appl. Phys. **51**, 355302 (2018)
10. A. le Febvrier, G. Abadias, N. van Nong, P. Eklund, Appl. Phys. Express **11**, 051003 (2018)
11. P. Eklund, S. Kerdsongpanya, B. Alling, J. Mater. Chem. C **4**, 3905 (2016)
12. J. He, T.M. Tritt, Science **357**, eaak9997 (2017)
13. Y. Du, J.Y. Xu, B. Paul, P. Eklund, Appl. Mater. Today **12**, 366 (2018)
14. J.H. Bahk, H.Y. Fang, K. Yazawa, A. Shakouri, J. Mater. Chem. C **3**, 10362 (2015)
15. O. Bubnova, Z.U. Khan, A. Malti, S. Braun, M. Fahlman, M. Berggren, X. Crispin, Nat. Mater. **10**, 429 (2011)
16. O. Bubnova, X. Crispin, Energy Environ. Sci. **5**, 9345 (2012)
17. J.L. Blackburn, A.J. Ferguson, C. Cho, Adv. Mater. **30**, 1704386 (2018)
18. W. Zhou, Q. Fan, Q. Zhang, L. Cai, K. Li, X. Gu, F. Yang, N. Zhang, Y. Wang, H. Liu, W. Zhou, S. Xie, Nat. Commun. **8**, 14886 (2017)
19. C. Yang, D. Souchay, M. Kneiß, M. Bogner, H. Wei, M. Lorenz, O. Oeckler, G. Benstetter, Y.Q. Fu, M. Grundmann, Nat. Commun. **8**, 16076 (2017)
20. B. Paul, J. Lu, P. Eklund, ACS Appl. Mater. Interfaces **9**, 25308 (2017)
21. B. Paul, E.M. Björk, A. Kumar, J. Lu, P. Eklund, ACS Appl. Energy Mater. **1**, 2261 (2018)
22. G.J. Snyder, E.S. Toberer, Nat. Mater. **7**, 105 (2008)
23. C.J. Vineis, A. Shakouri, A. Majumdar, M.G. Kanatzidis, Adv. Mater. **22**, 3970 (2010)
24. J.R. Sootsman, D.Y. Chung, M.G. Kanatzidis, Angew. Chim. Int. Ed. **48**, 8616 (2009)
25. L.D. Hicks, M.S. Dresselhaus, Phys. Rev. B **47**, 12727 (1993)
26. R. Venkatasubramanian, E. Siivola, T. Colpitts, B. O'Quinn, Nature **413**, 597 (2001)
27. C.B. Vining, Nat. Mater. **8**, 83 (2009)
28. H. Liu, X. Shi, F. Xu, L. Zhang, W. Zhang, L. Chen, Q. Li, C. Uher, T. Day, G.J. Snyder, Nat. Mater. **11**, 422 (2012)
29. R.J. Mehta, Y. Zheng, C. Karthik, B. Singh, R.W. Siegel, T. Borca-Tasciuc, G. Ramanath, Nat. Mater. **11**, 233 (2012)
30. G.D. Mahan, J.O. Sofo, Proc. Natl. Acad. Sci. U. S. A. **93**, 7436 (1996)
31. Z.Y. Fan, H.Q. Wang, J.-C. Zheng, J. Appl. Phys. **109**, 073713 (2011)
32. J.E. Sundgren, Thin Solid Films **21**, 128 (1985)
33. L. Hultman, Vacuum **57**, 1 (2000)
34. P.H. Mayrhofer, C. Mitterer, L. Hultman, H. Clemens, Prog. Mater. Sci. **51**, 1032 (2006)
35. R.S. Ningthoujam, N.S. Gajbhiye, Prog. Mater. Sci. **70**, 50 (2015)
36. L.M. Corliss, N. Elliott, J.M. Hastings, Phys. Rev. **117**, 929 (1960)
37. A. Herwadkar, W.R.L. Lambrecht, Phys. Rev. B **79**, 035125 (2009)
38. B. Alling, Phys. Rev. B **82**, 054408 (2010)
39. D. Gall, I. Petrov, J.E. Greene, J. Appl. Phys. **89**, 401 (2001)
40. L.E. Toth, *Transition Metal Carbides and Nitrides* (Academic Press, New York, 1971)
41. S.W. King, R.F. Davis, R.J. Nemanich, J. Vac. Sci. Technol. A **32**, 061504 (2014)
42. J.P. Dismukes, W.M. Yim, J.J. Tietjen, R.E. Novak, J. Cryst. Growth **9**, 295 (1971)
43. J.P. Dismukes, W.M. Yim, J.J. Tietjen, R.E. Novak, RCA Rev. **31**, 680 (1970)
44. G. Travaglini, F. Marabelli, R. Monnier, E. Kaldis, P. Wachter, Phys. Rev. B **34**, 3876 (1986)
45. W.R.L. Lambrecht, Phys. Rev. B **62**, 13538 (2000)
46. C. Stampfl, W. Mannstadt, R. Asahi, A.J. Freeman, Phys. Rev. B **63**, 155106 (2001)
47. D. Gall, M. Städele, K. Järrendahl, I. Petrov, P. Desjardins, R.T. Haasch, T.Y. Lee, J.E. Greene, Phys. Rev. B **63**, 125119 (2001)
48. M.A. Moram, Z.H. Barber, C.J. Humphreys, Thin Solid Films **516**, 8569 (2008)
49. R. Deng, B.D. Ozsdolay, P.Y. Zheng, S.V. Khare, D. Gall, Phys. Rev. B **91**, 045104 (2015)
50. D. Gall, I. Petrov, N. Hellgren, L. Hultman, J.E. Sundgren, J.E. Greene, J. Appl. Phys. **84**, 6034 (1998)
51. D. Gall, I. Petrov, P. Desjardins, J.E. Greene, J. Appl. Phys. **86**, 5524 (1999)

52. J.M. Gregoire, S.D. Kirby, G.E. Scopelianos, F.H. Lee, R.B. van Dover, J. Appl. Phys. **104**, 074913 (2008)
53. Y. Oshima, G.E. Villora, K. Shimamura, J. Appl. Phys. **115**, 153508 (2014)
54. H.A. Al-Brithen, A.R. Smith, D. Gall, Phys. Rev. B **70**, 045303 (2004)
55. T.D. Moustakas, R.J. Molnar, J.P. Dismukes, Proc. Electrochem. Soc. **96**(11), 197 (1996)
56. H.A.H. Al-Brithen, E.M. Trifan, D.C. Ingram, A.R. Smith, D. Gall, J. Cryst. Growth **242**, 345 (2002)
57. H. Al-Brithen, A.R. Smith, Appl. Phys. Lett. **77**, 2485 (2000)
58. M.A. Moram, T.B. Joyce, P.R. Chalker, Z.H. Barber, C.J. Humphreys, Appl. Surf. Sci. **252**, 8385 (2006)
59. M.A. Moram, S.V. Novikov, A.J. Kent, C. Norenberg, C.T. Foxon, C.J. Humphreys, J. Cryst. Growth **310**, 2746 (2008)
60. J.R. Sootsman et al., Angew. Chem. Int. Ed. **47**, 8618 (2008)
61. V. Rawat, Y.K. Koh, D.G. Cahill, T.D. Sands, J. Appl. Phys. **105**, 024909 (2009)
62. R. Amatya, R.J. Ram, J. Electron. Mater. **41**, 1011 (2012)
63. S. Kerdsongpanya, B. Alling, P. Eklund, Phys. Rev. B **86**, 195140 (2012)
64. M.G. Moreno-Armenta, G. Soto, Comput. Mater. Sci. **40**, 275 (2007)
65. H.C. Barshilia, N. Selvakumar, B. Deepthi, K.S. Rajam, Surf. Coat. Technol. **201**, 2193 (2006)
66. G.G. Fuentes, R. Rodriguez, J.C. Avelar-Batista, J. Housden, F. Montalá, L.J. Carreras, A.B. Cristóbal, J.J. Damborenea, T.J. Tate, J. Mater. Process. Technol. **167**, 415 (2005)
67. M.L. Kuruppu, G. Negrea, I.P. Ivanov, S.L. Rohde, J. Vac. Sci. Technol. A **16**, 1949 (1998)
68. T. Polcar, T. Kubart, R. Novák, L. Kopecký, P. Široký, Surf. Coat. Technol. **193**, 192 (2005)
69. T. Polcar, N.M.G. Parreira, R. Novák, Surf. Coat. Technol. **201**, 5228 (2007)
70. J. Lin, W.D. Sproul, J.J. Moore, Mater. Lett. **89**, 55 (2012)
71. J. Lin, W.D. Sproul, J.J. Moore, Surf. Coat. Technol. **206**, 2474 (2012)
72. D. Gall, C.-S. Shin, T. Spila, M. Odén, M.J.H. Senna, J.E. Greene, I. Petrov, J. Appl. Phys. **91**, 3589 (2002)
73. C. Petrogalli, L. Montesano, M. Gelfi, G.M. La Vecchia, L. Solazzi, Surf. Coat. Technol. **258**, 878 (2014)
74. J. Lin, N. Zhang, W.D. Sproul, J.J. Moore, Surf. Coat. Technol. **206**, 3283 (2012)
75. N. Beliardouh, K. Bouzid, C. Nouveau, B. Tlili, M. Walock, Tribol. Int. **82**, 443 (2015)
76. P.S. Herle, M. Hegde, N. Vasathacharya, S. Philip, M.R. Rao, T. Sripathi, J. Solid State Chem. **134**, 120 (1997)
77. O. Jankovský, D. Sedmidubský, V. Huber, P. Šimek, Z. Sofer, J. Eur. Ceram. Soc. **34**, 4131 (2014)
78. A. Filippetti, W.E. Pickett, B.M. Klein, Phys. Rev. B **59**, 7043 (1999)
79. A. Filippetti, N.A. Hill, Phys. Rev. Lett. **85**, 5166 (2000)
80. C. Constantin, M.B. Haider, D. Ingram, A.R. Smith, Appl. Phys. Lett. **85**, 6371 (2004)
81. P. Hones, M. Diserens, R. Sanjinés, F. Lévy, J. Vac. Sci. Technol. B **18**, 2851 (2000)
82. K. Inumaru, K. Koyama, N. Imo-oka, S. Yamanaka, Phys. Rev. B **75**, 054416 (2007)
83. X.Y. Zhang, J.S. Chawla, B.M. Howe, D. Gall, Phys. Rev. B **83**, 165205 (2011)
84. C.X. Quintela, F. Rivadulla, J. Rivas, Appl. Phys. Lett. **94**, 152103 (2009)
85. X.Y. Zhang, J.S. Chawla, R.P. Deng, D. Gall, Phys. Rev. B **84**, 073101 (2011)
86. D. Gall, C.-S. Shin, R.T. Haasch, I. Petrov, J.E. Greene, J. Appl. Phys. **91**, 5882 (2002)
87. P. Hohenberg, W. Kohn, Phys. Rev. **136**, B864 (1964)
88. W. Kohn, L.J. Sham, Phys. Rev. **140**, A1133 (1965)
89. J.P. Perdew, K. Burke, M. Ernzerhof, Phys. Rev. Lett. **77**, 3865 (1996)
90. A. Mei, O. Hellman, N. Wireklint, C.M. Schlepütz, D.G. Sandgiovanni, B. Alling, A. Rockett, L. Hutlman, I. Petrov, J.E. Greene, Phys. Rev. B **91**, 054101 (2015)
91. A.S. Botana, V. Tran, D. Pardo, D. Baldomir, P. Blaha, Phys. Rev. B **85**, 235118 (2012)
92. B. Alling, T. Marten, I.A. Abrikosov, Phys. Rev. B **82**, 184430 (2010)

93. F. Rivadulla, M. Banobre-Lopez, C.X. Quintela, A. Pineiro, V. Pardo, D. Baldomir, M.A. Lopez-Quintela, J. Rivas, C.A. Ramos, H. Salva, J.-S. Zhou, J.B. Goodenough, Nat. Mater. **8**, 947 (2009)
94. B. Alling, T. Marten, I.A. Abrikosov, Nat. Mater. **9**, 283 (2010)
95. S. Wang, X. Yu, J. Zhang, M. Chen, J. Zhu, L. Wang, D. He, Z. Lin, R. Zhang, K. Leinenweber, Y. Zhao, Phys. Rev. B **86**, 064111 (2012)
96. P. Steneteg, B. Alling, I.A. Abrikosov, Phys. Rev. B **85**, 144404 (2012)
97. B. Alling, L. Hultberg, L. Hultman, I.A. Abrikosov, Appl. Phys. Lett. **102**, 031910 (2013)
98. N. Shulumba, B. Alling, O. Hellman, E. Mozafari, P. Steneteg, M. Odén, I.A. Abrikosov, Phys. Rev. B **89**, 174108 (2014)
99. L. Zhou, F. Körmann, D. Holec, M. Bartosik, B. Grabowski, J. Neugebauer, P.H. Mayrhofer, Phys. Rev. B **90**, 184102 (2014)
100. I. Stockem, A. Bergman, A. Glensk, T. Hickel, F. Körmann, B. Grabowski, J. Neugebauer, B. Alling, Phys. Rev. Lett. **121**, 125902 (2018)
101. M. Dahlqvist, B. Alling, J. Rosén, J. Appl. Phys. **113**, 216103 (2013)
102. P. Eklund, J. Rosen, P.O.Å. Persson, J. Phys. D. Appl. Phys. **50**, 113001 (2017)
103. E. Mozafari, B. Alling, P. Steneteg, I.A. Abrikosov, Phys. Rev. B **91**, 094101 (2015)
104. P.-W. Ma, S.L. Dudarev, Phys. Rev. B **91**, 054420 (2015)
105. A. Lindmaa, R. Lizarraga, E. Holmström, I.A. Abrikosov, B. Alling, Phys. Rev. B **88**, 054414 (2013)
106. B. Alling, T. Marten, I.A. Abrikopsov, A. Karimi, J. Appl. Phys. **102**, 044314 (2007)
107. A. Zunger, S.-H. Wei, L.G. Ferreira, J.E. Bernard, Phys. Rev. Lett. **65**, 353 (1990)
108. B. Alling, A.V. Ruban, A. Karimi, O.E. Peil, S.I. Simak, L. Hultman, I.A. Abrikosov, Phys. Rev. B **75**, 045123 (2007)
109. F. Tasnádi, I.A. Abrikosov, I. Katardjiev, Appl. Phys. Lett. **94**, 151911 (2009)
110. F. Tasnádi, B. Alling, C. Höglund, G. Wingqvist, J. Birch, L. Hultman, I.A. Abrikosov, Phys. Rev. Lett. **104**, 137601 (2010)
111. S. Kerdsongpanya, B. Alling, P. Eklund, J. Appl. Phys. **114**, 073512 (2013)
112. L. Zhou, D. Holec, P.H. Mayrhofer, J. Phys. D **46**, 365301 (2013)
113. B. Alling, A.V. Ruban, A. Karimi, L. Hultman, I.A. Abrikosov, Phys. Rev. B **83**, 104203 (2011)
114. N. Shulumba, O. Hellman, Z. Raza, J. Barriero, B. Alling, F. Mücklich, I.A. Abrikosov, M. Odén, Phys. Rev. Lett. **117**, 205502b (2016)
115. F. Bernandini, V. Fiorentini, D. Vanderbilt, Phys. Rev. B **56**, R10024 (1997)
116. F. Tasnádi, M. Odén, I.A. Abrikosov, Phys. Rev. B **85**, 144112 (2012)
117. G.K.H. Madsen, D.J. Singh, Comput. Phys. Commun. **175**, 67 (2006)
118. K. Kutorasiński, J. Tobola, S. Kaprzyk, Phys. Rev. B **87**, 195205 (2013)
119. A. Faghaninia, J.W. Ager III, C.S. Lo, Phys. Rev. B **91**, 235123 (2015)
120. B. Qiu, Z. Tian, A. Vallabhaneni, B. Liao, J.M. Mendoza, O.D. Restrepo, X. Ruan, G. Chen, EPL **109**, 57006 (2015)
121. J. Zhou, B. Liao, B. Qiu, S. Huberman, K. Esfarjani, M.S. Dresselhaus, G. Chen, Proc. Natl. Acad. Sci. U. S. A. **112**, 14777 (2015)
122. S. Kerdsongpanya, B. Sun, F. Eriksson, J. Jensen, J. Lu, Y.K. Koh, N. Van Nong, B. Balke, B. Alling, P. Eklund, J. Appl. Phys. **120**, 215103 (2016)
123. M.A. Caro, S. Zhang, T. Riekkinen, M. Ylilammi, M.A. Moram, O. Lopez-Acevedo, J. Molarius, T. Laurila, J. Phys. Condens. Matter. **27**, 245901 (2015)
124. M.A. Moram, S. Zhang, J. Mater. Chem. A **2**, 6042 (2014)
125. J.M. Gregoire, S.D. Kirby, M.E. Turk, R.B. van Dover, Thin Solid Films **517**, 1607 (2009)
126. B. Saha, G. Naik, V.P. Drachev, A. Boltassseva, E.E. Marinero, T.D. Sands, J. Appl. Phys. **114**, 063519 (2013)
127. B. Saha, M. Garbrecht, J.A. Perez-Taborda, M.H. Fawey, Y.R. Koh, A. Shakouri, M. Martin-Gonzalez, L. Hultman, T.D. Sands, Appl. Phys. Lett. **110**, 252104 (2017)
128. B. Saha, J.A. Perez-Taborda, J.H. Bahk, Y.R. Koh, A. Shakouri, M. Martin-Gonzalez, T.D. Sands, Phys. Rev. B **97**, 085301 (2018)

129. N. Tureson, M. Marteau, T. Cabioch, N. Van Nong, J. Jensen, J. Lu, G. Greczynski, D. Fournier, N. Singh, A. Soni, L. Belliard, P. Eklund, A. le Febvrier, Phys. Rev. B **98**(20), 205307 (2018)
130. A. Shakouri, Annu. Rev. Mater. Res. **41**, 399 (2011)
131. E.S. Toberer, L.L. Baranowski, C. Dames, Annu. Rev. Mater. Res. **42**, 179 (2012)
132. K. Biswas, J. He, Q. Zhang, G. Wang, C. Uher, V.P. Dravid, M.G. Kanatzidis, Nat. Chem. **3**, 160 (2011)
133. P. Eklund, M. Beckers, U. Jansson, H. Högberg, L. Hultman, Thin Solid Films **518**, 1851 (2010)
134. G. Hug, Phys. Rev. B **74**, 184113 (2006)
135. M. Zebarjadi, Z. Bian, R. Singh, A. Shakouri, R. Wortman, V. Rawat, T.D. Sands, J. Electron. Mater. **38**, 960 (2009)
136. R. Niewa, D.A. Zherebtsov, W. Schnelle, F.R. Wagner, Inorg. Chem. **43**, 6188 (2004)
137. J.L. Schroeder, B. Saha, M. Garbrecht, N. Schell, T.D. Sands, J. Birch, J. Mater. Sci. **50**, 3200 (2015)
138. B. Saha, S. Saber, G.V. Naik, A. Boltasseva, E.A. Stach, E.P. Kvam, T.D. Sands, Phys. Status Solidi B. **252**, 251 (2015)
139. B. Saha, S.K. Lawrence, J.L. Schroeder, J. Birch, D.F. Bahr, T.D. Sands, Appl. Phys. Lett. **105**, 151904 (2014)
140. B. Saha, Y.R. Koh, J. Comparan, S. Sadasivam, J.L. Schroeder, M. Garbrecht, A. Mohammed, J. Birch, T. Fisher, A. Shakouri, T.D. Sands, Phys. Rev. B **93**(4), 045311 (2016)
141. M. Garbrecht, L. Hultman, M.H. Fawey, T.D. Sands, B. Saha, J. Mater. Sci. **53**, 4001 (2018)
142. M. Garbrecht, B. Saha, J.L. Schroeder, L. Hultman, T.D. Sands, Sci. Rep. **7**, 46902 (2017)
143. M. Garbrecht, L. Hultman, T.D. Sands, B. Saha, Phys. Rev. Mater. **1**, 033402 (2017)
144. B. Saha, A. Shakouri, T.D. Sands, Appl. Phys. Rev. **5**, 021101 (2018)
145. B. Alling, Phys. Rev. B **89**, 085112 (2014)
146. C. Tholander, F. Tasnadi, I.A. Abrikosov, L. Hultman, J. Birch, B. Alling, Phys. Rev. B **92**, 174119 (2015)
147. Y. Iwazaki, T. Yokoyama, T. Nishihara, M. Ueda, Appl. Phys. Express **8**, 061501 (2015)

Chapter 7
Thermoelectric Modules Based on Oxide Thin Films

Paolo Mele, Shrikant Saini, and Edoardo Magnone

7.1 Introduction

A huge amount (66%) of the total energy obtained from natural fossil resources is lost every year as waste heat [1], in a wide range of temperatures (300~1000 K) and from a variety of sources like human body, hot springs, industrial processes, home heating, lighting, electronic apparatuses, utility pipelines, electrical substations, subway networks, and automotive exhaust tubes. It is mandatory to harvest this heat to improve energy efficiency, reduce the overall CO_2 emissions, and enable extensive adoption of battery-free and electric-grid-free devices, wearable electronics, and more. A highly promising method for energy recovery from waste heat is the utilization of thermoelectric (TE) effect, known since 1821 [2].

To obtain a TE harvester, a joint between n-type TE material and p-type TE material is needed, forming an n–p TE couple, then put one side in contact to the hot surface and cool the opposite side. In this way, electricity is generated at the cold side (Fig. 7.1a). To increase the amount of harvested heat, several couples are connected forming a TE module. This can be done traditionally by cutting n and p bars from sintered pellets [3], then assembling them on a ceramic plate (Fig. 7.1b)

P. Mele (✉)
SIT Research Laboratories, Omiya Campus, Shibaura Institute of Technology, Tokyo, Japan
e-mail: pmele@shibaura-it.ac.jp

S. Saini
Department of Mechanical and Control Engineering, Kyushu Institute of Technology,
Kitakyushu, Japan
e-mail: saini.shrikant456@mail.kyutech.jp

E. Magnone
Department of Chemistry and Biochemical Engineering, Dongguk University, Seoul,
Republic of Korea

© Springer Nature Switzerland AG 2019   139
P. Mele et al. (eds.), *Thermoelectric Thin Films*,
https://doi.org/10.1007/978-3-030-20043-5_7

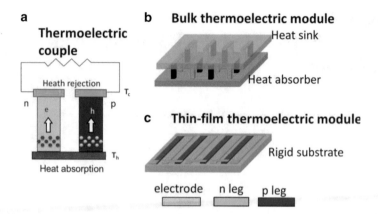

Fig. 7.1 (**a**) Schematic representation of a thermoelectric couple; (**b**) Combination of bars cut from bulk pellet forming a conventional bulk module; (**c**) Combination of several couples on a substrate forming a thin-film-based thermoelectric module. The three panels are not in the same scale

or—as demonstrated more recently—preparing then connecting n and p thin films on a rigid single crystal substrate [4] (Fig. 7.1c).

Conventional bulk modules (Fig. 7.1b) are used with heat flow parallel to the longer direction of the legs (i.e., perpendicular to their cross sections) in order to maximize the effect of the thermal gradient. The thin-film TE modules in principle can be utilized either in cross-plane or in-plane configuration. The cross-plane configuration is the same as the standard TE bulk module configuration, and the temperature gradient should flow perpendicularly to the substrate, i.e., across the cross-plane direction. This is the most relevant configuration for TE generator and cooling solutions but presents practical difficulties to be applied since the effect of the thermal gradient may be very small due to the typical thickness of the thin films (hundreds of nanometers). The in-plane configuration, where the temperature gradient arises parallel to the thin film and the substrate, is not that relevant for the generator or cooling solutions, but it can be very useful for TE sensor applications providing improved sensitivity and form factor in comparison with conventional temperature sensors.

TE modules are very attractive because they are capable of converting waste heat directly into electricity without moving parts. Consequently, research flourished and a lot of TE materials (i.e., alloys and oxides) were discovered. However, after a century since the discovery of the TE effect, the practical use of TE devices is quite limited.

There are three major obstacles to the wide diffusion of TE harvesters: toxicity and scarcity of elements constituting TE materials; low conversion efficiency and low stability of TE materials; mechanical rigidity of TE modules. TE modules based on oxide thin films may solve these issues in the near future, offering stability, environmental sustainability, improved efficiency, and flexibility. The scope of this chapter is to briefly review the current state of the art of TE modules based on oxide thin films.

The chapter is divided into three parts. In the first part, it is described how the TE thin-film modules are fitted to solve the aforementioned issues; in the second part literature survey is offered; in the last and third part, a conclusion and some perspectives are given.

7.2 The Promise of Oxide Thin Films Thermoelectric Modules

As reported in the literature, oxide thin films can be used to produce stable and environmentally friendly TE materials, surpassing the performance of the corresponding bulk oxides in the same interval of temperatures [5]. In addition, some preliminary results suggest that a significant improvement in their efficiency may be obtained through the addition with nano-sized artificial defects.

Let's consider in detail both aspects.

7.2.1 Use of Oxide Thin Films as Sustainable Thermoelectric Materials

Mainly the commercial TE modules are made by materials containing critical raw elements like Pb, Bi, Hg (toxic), Ag, Te, and Se (expensive). The consequent high prices of the modules and concerns related to environmental impact have confined their use to niche applications (for example, the power supply of space probe Cassini) and drastically limited their diffusion in the normal daily life.

This issue can be solved by using thermally stable, inexpensive, and sustainable TE oxide materials to substitute the conventional unstable, toxic, and expensive TE materials (BiTe, PbTe, AgSbGeTe, and so on). All of these oxides allow heat harvesting in a wider range of temperatures (Fig. 7.2). In the 1990s, extensive research on bulk oxide flourished with common effort focused to enhance the TE performance—defined through a dimensionless figure of merit (ZT)—by atomic substitutions and improved grain connection [6, 7]. However, the best TE performance of the currently available oxide materials (ZT $\sim$0.64 for n-type [8] and 0.74 for p-type [9] oxides at 1000 K) is not yet up to the level of the best conventional TE materials and requires to be improved. Furthermore, oxide thin films offer significant advantages in respect to bulks for developing TE: flexibility, rapid fabrication, control of defects at the nanoscale [10]. Several attempts have been tried on thin films of Al-doped ZnO (AZO) films fabricated by pulsed laser deposition (PLD) on several single crystals (i.e., $SrTiO_3$ and Al_2O_3) and amorphous (silica) substrates.

Independently on the substrate, films always show values of the figure of merit ZT in comparison with corresponding bulk AZO: for example, at $T = 600$ K, $(ZT)_{AZO\text{-}on\text{-}STO} = 0.03$ [5, 11].

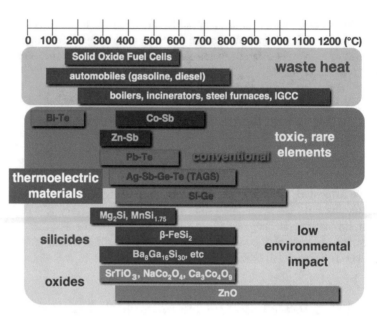

Fig. 7.2 Typical range of waste heat sources and operating temperatures of n- and p-type thermoelectric oxide materials. Reproduced with permission from [7]

The figure of merit (ZT) of a TE material is defined as

$$ZT = \left(\sigma S^2 \right) \times T / \left(\kappa_{el} + \kappa_{ph} \right) \tag{7.1}$$

where σ is the electrical conductivity, S is the Seebeck coefficient ($S = \Delta V/\Delta T$, being V the electrical voltage), κ is the total thermal conductivity ($\kappa = \kappa_{el} + \kappa_{ph}$, where κ_{el} and κ_{ph} are the electronic and phononic contribution, respectively), and T is the operating temperature.

The superior performance of films is due to their lower thermal conductivity: $\kappa_{\text{AZO-on-STO}}$ (300 K) = 6.5 W/m × K [5, 11], while κ_{BULK} (300 K) = 34 W/m × K.

In this first series of films, the grain boundaries can be considered as natural nanodefects for the enhanced scattering of phonons and consequent depression of κ respect to the bulk material. As a demonstration of this effect, the film on fused silica, showing additional grain boundaries at the seed layer on the substrate, had even lower thermal conductivity: κ_{silica} (300 K) = 4.89 W/m × K.

7.2.2 Enhancement of Conversion Efficiency by Addition of Controlled Nanodefects

Typically the commercially available TE modules have a figure of merit ZT lower than one (efficiency ≤10%). In general, metal alloys are unstable chemical compounds in air, decomposing at relatively low temperatures (usually 500 K or less). The parameters σ, S, and κ, however, are not independently tunable; for example, conventionally it is not possible to increase σ and reduce κ at the same time, since

$$\kappa_{el} = LT\sigma \tag{7.2}$$

with $L = 2.44 \times 10^{-8}$ W Ω K^{-2} (Lorentz number). On the other hand,

$$\kappa_{ph} = 1/3 \, Cv\Lambda \tag{7.3}$$

with C = heat capacity, v = speed of phonons, Λ = phonons mean free path.

The presence of nano-sized defects has the effect of scattering the phonons because their mean free path Λ is on the order of several nanometers; this will result in the reduction of k_{el}, (Eq. 7.3). According to Eq. (7.1), if κ is reduced to 1 W m^{-1} K^{-1} (without significantly depressing σ and S) by introduction of artificial nanodefects, the ZT value could reach 2.1 (efficiency >20%), at 800 K overcoming the typical values of bulk oxides (ZT = 0.75, maximum efficiency ~7%). According to Eqs. (7.1)–(7.3), it is mandatory to insert uniformly spaced nanodefects whose size is 20 nm or less, and density $10^{24} \sim 10^{25}$ m^{-3} (Fig. 7.3). Larger densities of nanoparticles may be detrimental for the current transport. Nanoparticles and nanovoids can be inserted in oxide bulk materials prepared by conventional solid-state synthesis techniques; however, the control of their size and distribution is quite difficult. On the contrary, by nanoengineering approach during the fabrication of oxide thin films, it is possible to control size and distribution of artificial nanodefects, as was already demonstrated for superconducting films [12, 13].

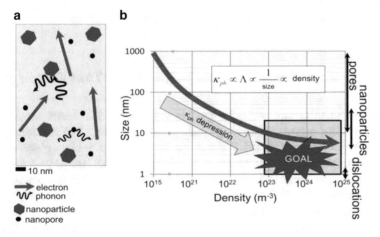

Fig. 7.3 (a) schematic representation of phonon scattering by different nanodefects inside a thermoelectric material; (b) size–density relationship for several kinds of nanodefects. Required size and density for efficient phonon scattering in nanostructured thin films is highlighted (pink area). Right-side panel is reproduced with permission from [10]

This kind of approach is still at a near stage for TE oxide films, but some artificial nanostructures were already introduced yielding enhancement of TE performance.

Based on the recent literature, it can be summarized in the following: (1) insertion of hydroquinone nanolayers in AZO films prepared by atomic layer deposition (ALD): κ_{ALD} (300 K) = 3.56 W/m × K [14]; (2) addition of poly(methyl methacrylate) (PMMA) particles to AZO films prepared by multi-beam multi-target matrix-assisted PLD (MBMT/MAPLE-PLD): κ_{MAPLE} (300 K) = 5.9 W/m × K [15]; (3) formation of nanopores in AZO films prepared by Mist-Chemical Vapor Deposition (Mist-CVD): κ_{porous} (300 K) = 0.60 W/m × K [16]; (4) dispersion of Al_2O_3 nanoparticulate in AZO films prepared by surface-modified target PLD: $\kappa_{nanoAl2O3}$ (300 K) = 3.98 W/m × K [17]. Reduction of κ is one-tenth or more respect to the bulk AZO value.

The common issues of all these approaches are the control of size as well as the distribution of nanodefects.

7.3   *State-of-the-art* of Oxide Thin Films Thermoelectric Modules

Recently, several reports on TE modules based on doped oxide thin films have been published. The most common arrangement presents parallel legs of n- and p-type oxides deposited by means of different techniques (PLD, ALD, sputtering). While different p-type oxides (i.e., CuO, $Ca_3Co_4O_9$, and $NaCoO_2$) have been chosen, constantly the n-type legs are constituted by M-doped ZnO where M is a metal ion. This is not surprising since Al-doped ZnO (M=Al) is the most studied TE oxide as thin film or bulk. In other configurations, hybrid modules are based on oxides and metallic thin-film legs and on uni-leg (only one kind of film is utilized).

7.3.1   *Modules Based on n- and p-Type Oxide Thin Film Legs*

Firstly, W. Somkhunthot and colleagues proposed a module based on the combination of four n-type Al-ZnO legs (called "AZO" in Fig. 7.4) and four p-type $Ca_3Co_4O_9$ (CCO) legs [18, 19]. The thin-film legs are prepared on a glass substrate of 1.0 mm thickness in dimension mm^2 by bipolar pulsed-DC magnetron sputtering in an argon atmosphere and then connected by Cu electrodes (Fig. 7.4). Each leg size is 3.0 mm width, 20.0 mm length, and 440 nm thickness. A hot plate (absorbed heat, T_H) was placed on a module to heat at 300–400 K, while a cold plate (released heat, T_C) was surrounded by air at room temperature. The temperature difference ($\Delta T = T_H - T_C$), open circuit voltage (V_O), and internal resistance (R_I) of a module were measured. It was found that the open circuit voltage increased with increasing temperature difference from $V_O = 3$ mV at $\Delta T = 5$ K ($T_H = 305$ K and $T_C = 300$ K)

Fig. 7.4 Test results on a thermoelectric module of n-Al-ZnO ("AZO") and p-Ca$_3$Co$_4$O$_9$ ("CCO") thin films: (**a**) the open circuit voltage and (**b**) the short circuit current as a function of temperature difference. Reproduced with permission from [19]

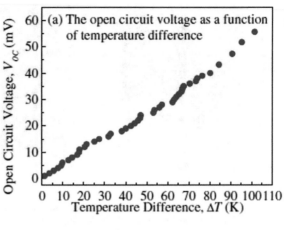

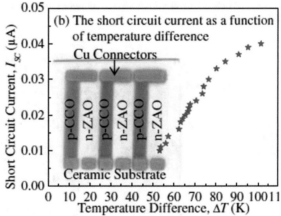

up to $V_O = 20$ mV at $\Delta T = 78$ K ($T_H = 393$ K and $T_C = 315$ K). The internal resistance of a module reached a value of 14.52 MΩ when $\Delta T = 78$ K.

The same group subsequently reported about a module based on the combination of three n-type AZO legs and three p-type NaCoO$_2$ (NCO) legs, prepared in the same way as their AZO/CCO thin-film module [20]. After the module test, the open circuit voltage increased with temperature up to 26.0 mV for a $\Delta T = 79.3$ K. However, the module was unable to produce useful electrical current due to its high internal series resistance contributed from the NCO films.

Zappa et al. [21] proposed a planar TE generator combining bundles of undoped ZnO and CuO nanowires in a series of five thermocouples on (2×2) cm^2 alumina substrate. At first, ZnO nanowires were fabricated by physical vapor deposition (PVD) technique in a tubular furnace, then using a shadow mask technique, Cu stripes were deposited via RF magnetron sputtering and finally the CuO nanowires were obtained via oxidation at 400 °C for 12 h in 80% oxygen/20% argon atmosphere (300 sccm flow). The module has five n-type AZO legs and five p-type

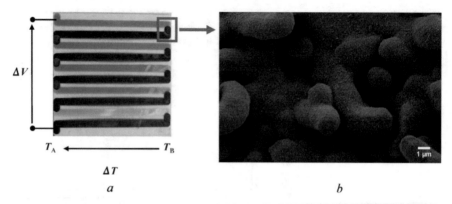

Fig. 7.5 (**a**) Optical image of fabricated planar thermoelectric device based on ZnO and CuO nanowires. (**b**) SEM picture of the ZnO–CuO junction area. Reproduced from [21] (open access article) under the terms of the Creative Commons Attribution License (http://creativecommons. org/licenses/by/2.0)

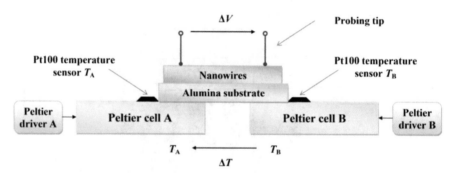

Fig. 7.6 Schematic diagram (side view) of the experimental setup for the measurement of the thermoelectric response of nanowires-based samples (not to scale)—Reproduced from [21] (open access article) under the terms of the Creative Commons Attribution License (http:// creativecommons.org/licenses/by/2.0)

CuO legs (Fig. 7.5). AZO film has typical Seebeck coefficient $S = 0.19$ mV/K and electrical conductivity $\sigma = 0.7$ S/m, while CuO has $S = +0.82$ mV/K and $\sigma = 2$ S/m.

The TE response of the module was measured as a function of the applied temperature difference ΔT using a purposely developed experimental setup based on two Peltier cells and two Pt-100 thermometers (Fig. 7.6) The Seebeck coefficient S of the entire planar device was measured, resulting in approximately 4 mV/K. The electrical resistance of the entire TE planar generation was about 9 MΩ.

The fabricated planar TE generators exhibit a maximum power $P_{max}/\Delta T^2$ of about 0.4 pW/K^2 (i.e., 102 pW at the maximum $\Delta T = 16$ K), comparable with values reported in similar generators based on Si and Si–Ge nanowire arrays [22].

Saini et al. [23] described planar modules based on five legs of n-type AZO and five legs of p-type CCO thin films on three kinds of substrates (i.e., Al$_2$O$_3$,

SrTiO$_3$, and silica glass). At 600 K, depending on the substrate, typical performances of standalone AZO films fabricated by PLD are $\sigma = 291\sim923$ S/cm, $S = -111\sim126$ μV/K, and $\kappa = 4.89\sim6.5$ W/m K [10, 11]; of standalone CCO films are $\sigma = 1.23\sim18$ S/cm, $S = +158\sim186$ μV/K, while κ was not measured [23].

The module legs were fabricated on 1 cm^2 substrate by pulsed laser deposition (PLD) technique using Nd:YAG laser (266 nm). At first, the laser was focused on CCO target, and p-type legs were deposited in an oxygen atmosphere at 650 °C on 10×10 mm substrate by superimposing a custom nickel mask. Then, AZO target was moved under the laser beam and ablated at 400 °C to fabricate n-type legs after shifting the custom Ni mask. Gold electrodes were sputtered using a custom Ni mask at room temperature after completion of PLD routes in order to achieve the electrical connection of the p–n couples. Performance of modules was evaluated using ad hoc customized system where the module was set vertically in between alumina (heat source; however heat is generated by hot plate) and aluminum nitride (heat sink) sheet. A schematic drawn and a picture of the module and evaluation setup, along with an optical image of the module, are reported in Fig. 7.7.

The power generation characteristics of on-chip TE module are shown in Fig. 7.8 for the case of maximum obtained temperature difference ($\Delta T = T_h - T_c$) between the heat source (T_h) and sink (T_c). The maximum output power (P_{max}) increases with an increase in ΔT and the highest output power is achieved while keeping higher ΔT.

Overall, the best performance of on-chip TE was obtained on Al$_2$O$_3$: $P_{max} = 29.9$ pW.

Mondarte et al. [24] proposed TE module based on n-type AZO and p-type N-doped Cu$_x$O. Eight n-type legs and eight p-type legs were deposited on (1.27×0.64) cm^2 glass substrate by spray pyrolysis and connected by Ag electrodes as schematically illustrated in Fig. 7.9. Standalone AZO at 300 K had $S = 0.448$ mV/K, $\sigma = 2440$ S/m, and PF $= \sigma S^2 = 4.8710^{-4}$ W/m K^2, while N-doped Cu$_x$O had $S = -1.002$ mV/K, $\sigma = 53$ S/m, and PF $= \sigma S^2 = 5.54 \times 10^{-5}$ W/m K^2.

The TE device based on the deposited TE materials in glass substrates generated a maximum of 32.8 mV when subjected to $\Delta T \sim28$ K.

7.3.2 Hybrid Modules Based on Oxide and Metallic Legs

Park et al. [25] reported on a hybrid module, made by four n-type Al$_2$O$_3$/ZnO (AO/ZnO) superlattice legs and four p-type Bi$_{0.5}$Sb$_{1.5}$Te$_3$ (BST) legs. The AO/ZnO films are grown on Si/SiO$_2$ substrates by atomic layer deposition (ALD) at 250 °C in a home-made reactor, and the balance of AO and ZnO layers was chosen to obtain 2% atomic concentration of Al in ZnO. The AO/ZnO superlattice films present $S = -62.4$ μV K^{-1}, $\sigma = 113$ S/cm, and $\kappa = 0.96$ W m^{-1} K^{-1} at 300 K, with ZT $= 0.014$, twice than a conventional AO/ZnO film. The p-BST films are grown

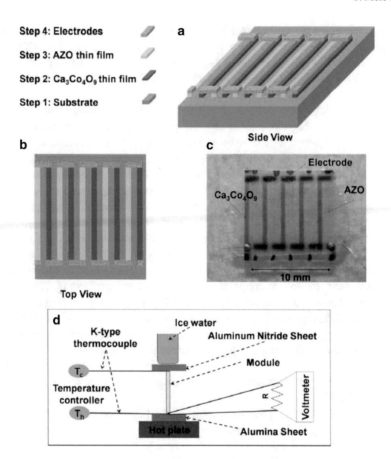

Fig. 7.7 (**a**) 3D sketch of the thin-film module based on AZO and Ca₃Co₄O₉ with sequence of thin films and electrodes deposition; (**b**) schematic top view of the module and (**c**) photograph of the module after fabrication of thin-films legs and gold contacts; (**d**) schematic of module testing setup with a load resistor (*R*). Reproduced with permission from [23]

by radio-frequency (RF) sputtering at room temperature, followed by annealing at 200 °C. The p-BST presents typical values of $S = 390\ \mu\text{V K}^{-1}$, $\sigma = 129$ S/cm, and power factor $= \sigma S^2 = 2.7 \times 10^{-3}$ W/K² at room *T*.

The fabrication of the module on a (20 × 15 mm) Si/SiO₂ substrate is done by alternating ALD deposition of n-type superlattice, photolithography to form the n-type legs, then sputtering deposition of p-type film followed by photolithography to form the p-type legs, and finally deposition of Ti/Au contacts by sputtering (Fig. 7.10). The separation between legs is 1 mm and the length of each leg is 8 mm.

The output voltage of the TE energy generator was measured as a function of the temperature difference applied between a hot plate and a cold junction (aluminum

Fig. 7.8 Output power of thin-film thermoelectric modules fabricated on the three different substrates at Th = 300 °C. Reproduced with permission from [23]

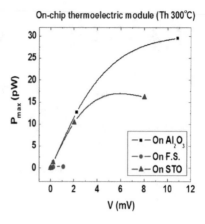

Fig. 7.9 Thermoelectric device based on n-type Al-doped ZnO (AZO) and p-type N-doped Cu_xO. Reproduced with permission from [24]

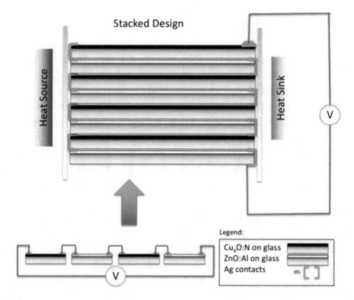

block) of the thin-film TE energy generators, and the maximum output power was estimated at room temperature by using a variable resistor.

The output power of the 100-nm-thick n-AO/ZnO superlattice film/p-BST TE energy generator reported in Fig. 7.11a was determined to be ~1.0 nW at a temperature difference of 80 K, corresponding to a significant improvement of 130% compared to the 100-nm-thick AO/ZnO conventional film/p-BST and 220% compared to n-BT/p-BST film generators (see Fig. 7.11b).

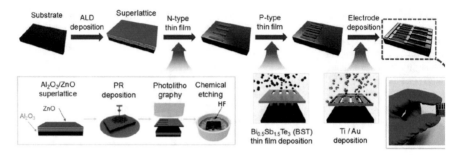

Fig. 7.10 Schematic of the fabrication process for an n-AO/ZnO superlattice film/p-BST TE energy generator consisting of four pairs of 100-nm-thick n-AO/ZnO superlattice films and 100-nm-thick p-BST thin-film legs on a Si substrate. Reproduced with permission from [25]

7.3.3 Uni-Leg Modules

In uni-leg modules, only one kind of legs, n-type or p-type, is constituted by oxides, while and the other kind of legs is made by reference electrode (Au, Cu, or alloy). Schematic representation of the uni-leg module is given in Fig. 7.12.

Rudez et al. [26] proposed the development of thick-film TE microgenerators based on n-type $(ZnO)_5In_2O_3$(Z5I) and p-type $Ca_3Co_4O_9$(CCO) uni-legs, while the reference electrodes are made by Pt or Pd/Ag ink in the case of Z5I thermopile and by Pd/Ag ink in the case of CCO. The Z5I and CCO legs are screen-printed (from a mixture of oxide powder, an organic binder, and wetting agent) as a rectangular structure (see Fig. 7.13) on alumina substrates and the modules are made by ten legs.

The overall performance of the devices was determined by measuring with a home-made device the electrical conductivity and the Seebeck coefficient of the Z5I and CCO thick-film thermopiles in the temperature range from room temperature to 500 °C. The CCO-based thermopile has a Seebeck coefficient of about 156 μV/K at 300 °C and resistivity as low as 13.7 mΩ cm, measured at 500 °C. The power factor calculated for a single leg was 1.6×10^4 W/mK^2. In the case of Z5I, the Seebeck coefficient and resistivity measured at 500 °C were 176 μV/K and 530 mΩ cm, respectively, with a calculated power factor of 1.4×10^6 W/mK^2.

Sinnarasa et al. [27] prepared a uni-leg module made with three legs of 100-nm-thick CuCrO$_2$:3% Mg film deposited by RF magnetron sputtering on fused silica, connected with Au electrodes. In order to characterize the module, one side of the module was heated and the other side was left untouched under air as shown in the inset of Fig. 7.14b. The applied temperature at the hot side and the measured temperature at the cold side of the module were plotted in Fig. 7.14a. Figure 7.14b

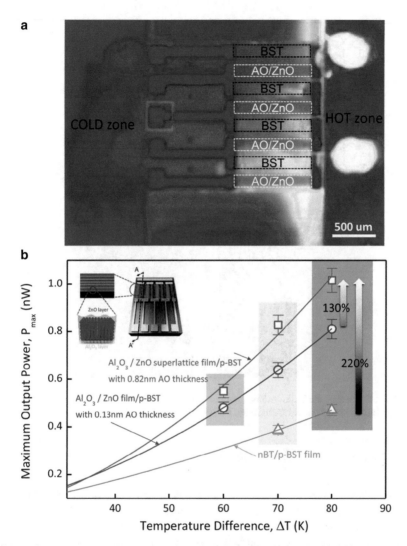

Fig. 7.11 (**a**) Thermography image of the n-AO/ZnO and p-BST TE power generator captured using an infrared (IR) camera during the measurements. (**b**) Comparison of the output power as a function of temperature differences up to 80 K for n-AO/ZnO superlattice/p-BST, n-AO/ZnO film/p-BST, and n-BT/p-BST thin-film-based TE energy generators. Reproduced with permission from [25]

shows the maximum electrical power generated by the three legs TE module when the hot side temperature is increased. In particular, it can be noted that it reached 10.6 nW when a temperature of 220 °C is applied at the hot side.

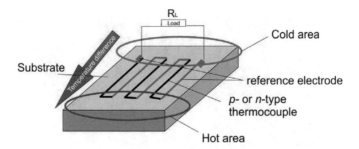

Fig. 7.12 Schematic view of a uni-leg thermoelectric module. Reproduced with permission from [26]

Fig. 7.13 Fabricated thick-film microgenerator from Ca_3CoO_9 (Ca349) (left) and $(ZnO)_5In_2O_3$ (Z5I) (right). Reproduced with permission from [26]

7.4 Summary and Perspective

In this chapter, thermoelectric (TE) modules based on oxide thin films were reviewed. Thin films prepared by several techniques (PLD, ALD, magnetron sputtering, and so on) on several kinds of substrates were considered, and TE properties were discussed in relation to structure and morphology. Summarizing the experimental data presented in this chapter (see Table 7.1), it is apparent that the output power generated by oxide thin films modules (in the order of pico- or nano-watts) is quite low in comparison with typical values of bulk modules (in the order of milli-watt).

Improvement in the output power of oxide thin films modules is expected according to the following guidelines:

1. Enhancement of the TE performance of the oxide thin films (lower κ and higher σ) by controlled addition of artificial nanodefects;
2. Implementation of a higher number of n–p couples;
3. Use of substrates of a larger area.

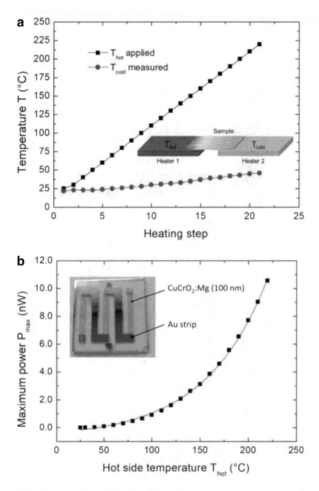

Fig. 7.14 (**a**) Applied temperature at the hot side and measured temperature at the cold side as a function of the heating step. Inset: Schematic representation of the setup. (**b**) The maximum power as a function of the hot side temperature. Inset: Photo of the uni-leg module. Reproduced with permission from [27]

As a final consideration, all the TE devices described in this chapter are based on thin films deposited on a crystal or glass substrate, then electrically connected to obtain TE modules. This means that typically the oxide thin films TE modules are rigid and flat. However, heat might be harvested from surfaces that may assume a variety of shapes (i.e., curved, zig-zag, irregular, and so on).

TE devices adaptable to curved surfaces can be obtained by conventional thin film oxides deposition techniques on flexible substrates (Fig. 7.15). n- and p-type elements adapt perfectly to the surface of the object, and heat can be harvested efficiently. Flexible TE modules, limited to the harvesting of body heat, have been prepared using conventional TE materials (see for example [28]).

Table 7.1 Characteristics of thermoelectric oxide thin films modules available in literature

n-Type	p-Type	Contacts	Number of legs (n + p)	Deposition technique	Size of a single leg	Sub.	Sub. size (mm L × mm W)	ΔT (°C)	Output power	Ref.
Al-ZnO	$Ca_3Co_4O_9$	Cu	4 + 4	DC sputtering	20 mm × 3 mm × 440 nm t	Glass	25.0 × 50.0	78	N/A	[18] [19]
Al-ZnO	$NaCoO_2$	Cu	3 + 3	DC sputtering	20 mm × 3 mm × 600 nm t	Al_2O_3	25.0 × 50.0	79.3	N/A	[20]
ZnO	CuO	Direct overlap	5 + 5	PVD	20 mm × 1 mm × 1000 nm t	Al_2O_3	20.0 × 20.0	16	102pW[a]	[21]
Al-ZnO	$Ca_3Co_4O_9$	Au	5 + 5	PLD	8 mm × 3 mm × 300 nm t	Glass	10.0 × 10.0	230	0.3 pW	[23]
Al-ZnO	$Ca_3Co_4O_9$	Au	5 + 5	PLD	8 mm × 3 mm × 300 nm t	$SrTiO_3$	10.0 × 10.0	230	16 pW	[23]
Al-ZnO	$Ca_3Co_4O_9$	Au	5 + 5	PLD	8 mm × 3 mm × 300 nm t	Al_2O_3	10.0 × 10.0	230	29.9 pW	[23]
Al-ZnO	$N-Cu_x$	Ag	8 + 8	Spray pyrolysis	N/A	Glass	12.7 × 6.4	28	N/A	[24]
Al_2O_3/ZnO	$Bi_{0.5}Sb_{1.5}Te_3$	Ti/Au	4 + 4	ALD (n)/ sputtering (p)	8 mm × 3 mm × 200 nm t	Si/SiO_2	20.0 × 15.0	80	1 nW	[25]
$(ZnO)_5In_2O_3$	Pt	Pt	10 + 10	Screen printing	N/A	N/A	260 × 176	N/A	N/A	[26][b]
Pd/Ag	$Ca_3Co_4O_9$	Pd/Ag	10 + 10	Screen printing	N/A	N/A	260 × 176	N/A	N/A	[26][b]
Au	$CuCrO_2$:3% Mg	Au	3 + 3	RF sputtering	100 nm t[c]	Glass	25.0 × 25.0	170	10.6 nW	[27][b]

[a]Calculated from relation $p = 0.4$ pW/ΔT^2 reported in [21]
[b]Uni-leg module
[c]Length and width of legs were not reported

Fig. 7.15 Schematic design of flexible TE module stuck on a pipe transporting a hot fluid

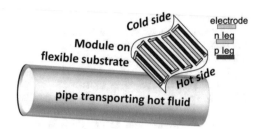

Flexible, ubiquitous, sustainable TE modules based on efficient oxide materials are predicted as a breakthrough for local heat harvesting in a wide range of temperatures.

References

1. BP Statistical Review of World Energy (2019); url (consulted 17 06 2019), https://www.bp.com/content/dam/bp/business-sites/en/global/corporate/pdfs/energy-economics/statistical-review/bp-stats-review-2019-full-report.pdf
2. T. Seebeck, Ann. Phys. **82**, 133 (1821)
3. P. Mele et al., Metals Mater. Int. **20**, 389S (2014)
4. S. Saini, P. Mele, et al., Energy Conv. Manag. **114**, 251 (2016)
5. P. Mele, S. Saini, H. Honda, et al., Appl. Phys. Lett. **253903**, 102 (2013)
6. J.W. Fergus, J. Eur. Ceram. Soc. **32**, 525 (2012)
7. *Nanostructured Oxide thermoelectric Materials with Enhanced Phonon Scattering*, ed. by M. Ohtaki. Chapter 8 in Oxide Thin Films, Multilayers and Nanocomposites (Springer, Berlin, 2015). https://doi.org/10.1007/978-3-319-14478-8
8. M. Ohtaki et al., J. Electron. Mater. **38**, 1234 (2009)
9. S. Saini et al., Sci. Rep. **7**, 44621 (2017)
10. *Nanostructured thin films of thermoelectric oxides*, ed. by P. Mele, Chapter 8 in Oxide Thin Films, Multilayers and Nanocomposites (Springer, Berlin, 2015). https://doi.org/10.1007/978-3-319-14478-8
11. S. Saini, P. Mele, et al., Jpn. J. Appl. Phys. **53**, 060306 (2014)
12. P. Mele et al., Supercond. Sci. Technol. **20**, 616 (2007)
13. M. Miura, B. Maiorov, P. Mele, et al., NPG Asia Mater. (2017). https://doi.org/10.1038/am.2017.197
14. T. Tynell, P. Mele, M. Karppinen, et al., J. Mater. Chem. A **2**, 12150 (2014)
15. A. Darwish, P. Mele et al., chapter 10 in Laser Ablation - From Fundamentals to Applications "INTECH 2018"
16. S. Saini, P. Mele, T. Oyake, J. Shiomi, J.-P. Niemela, M. Karppinen, K. Miyazaki, C. Li, T. Kawaharamura, A. Ichinose, L. Molina-Luna, Thin Solid Films **685**, 180 (2019). https://doi.org/10.1016/j.tsf.2019.06.010
17. P. Mele, S. Saini, M.I. Adam, S.J. Singh, et al., in preparation
18. W. Somkhunthot, N. Pimpabute, T. Seetawan, Adv. Mater. Res. **622**(623), 726 (2013)
19. W. Somkhunthot, N. Pimpabute, A. Vora-ud, T. Seetawan, T. Burinprakhon, Energy Procedia **61**, 795 (2014)
20. W. Somkhunthot, N. Pimpabute, A. Vora-ud, T. Seetawan, T. Burinprakhon Adv, Mater. Res. **931-932**, 386 (2014)

21. D. Zappa, S. Dalola, G. Faglia, E. Comini, M. Ferroni, C. Soldano, V. Ferrari, G. Sberveglieri, Beilstein J. Nanotechnol. **5**, 927 (2014)
22. B. Xu, C. Lia, M. Myronov, K. Fobelets, Solid State Electron. **83**, 107 (2013)
23. S. Saini, P. Mele, K. Miyazaki, A. Tiwari, Energ. Conv. Manag **114**, 251 (2016)
24. E.A. Mondarte, V. Copa, A. Tuico, C.J. Vergara, E. Estacio, A. Salvador, A. Somintac, Mat. Sci. Semicond. Proc. **45**, 27 (2016)
25. N.-W. Park, J.-Y. Ahn, T.-H. Park, J.-H. Lee, W.-Y. Lee, K. Cho, Y.-G. Yoon, C.-J. Choi, J.-S. Park, S.-K. Lee, Nanoscale **9**, 7027 (2017)
26. R. Rudež, P. Markowski, M. Presečnik, M. Košir, A. Dziedzic, S. Bernik, Ceram. Int. **41**, 13201 (2015)
27. I. Sinnarasa, Y. Thimont, L. Presmanes, A. Barnabé, P. Tailhades, J. Appl. Phys. **124**, 165306 (2018)
28. L. Francioso, C. De Pascali, I. Farella, C. Martucci, P. Cretì, P. Siciliano, A. Perrone, J. Power Sources **196**, 3239 (2011)

Chapter 8
Thermoelectric Properties of Metal Chalcogenides Nanosheets and Nanofilms Grown by Chemical and Physical Routes

Ananya Banik, Suresh Perumal, and Kanishka Biswas

8.1 Introduction

To alleviate the vivid surge of the worldwide energy crisis, furnishing renewable, sustainable and environment-friendly energy source holds precedence not only to the scientific community yet added to the overall populace all in all [1–3]. Thermoelectric (TE) materials are gaining attention in this area because of their ability in waste heat to electricity conversion [1]. The efficiency of any thermoelectric material is governed by the thermoelectric figure of merit, zT, which is the function of electrical conductivity (σ), Seebeck coefficient (S), total thermal conductivity (κ_{total}) and temperature (T) via the expression, $zT = \sigma S^2 T / \kappa_{total}$ [3]. The quantity σS^2 is called power factor (PF). κ_{total} of a material depends on electronic (κ_{el}) and lattice (κ_{lat}) thermal conductivity ($\kappa_{total} = \kappa_{el} + \kappa_{lat}$). A high performance thermoelectric material should have all together high electrical conductivity (σ) to minimize the internal Joule heating, a large Seebeck coefficient

A. Banik
New Chemistry Unit, Jawaharlal Nehru Centre for Advanced Scientific Research (JNCASR), Bangalore, India

S. Perumal
Department of Physics and Nanotechnology, SRM Institute of Science and Technology, Chennai, Tamil Nadu, India

K. Biswas (✉)
New Chemistry Unit, Jawaharlal Nehru Centre for Advanced Scientific Research (JNCASR), Bangalore, India

School of Advanced Materials, Jawaharlal Nehru Centre for Advanced Scientific Research (JNCASR), Bangalore, India

International Centre for Materials Science, Jawaharlal Nehru Centre for Advanced Scientific Research (JNCASR), Bangalore, India
e-mail: kanishka@jncasr.ac.in

© Springer Nature Switzerland AG 2019
P. Mele et al. (eds.), *Thermoelectric Thin Films*,
https://doi.org/10.1007/978-3-030-20043-5_8

(S) to generate high voltage and low thermal conductivity (κ_{total}) to maintain the temperature gradient. However, the major challenge relies on the improvement of the material's thermoelectric performance involving the optimization of these three interdependent material properties [2].

During the last decade various novel strategies have been introduced to decouple these interrelated physical parameters [2, 4, 5]. Approaches to improve the power factor (σS^2) include band convergence [6, 7], minority carrier filtering [8], quantum confinement effect [9–11] and formation of sharp impurity states close to Fermi level to increase density of states (DOS) [12]. To block the thermal transport for maintaining temperature difference across a given material/device, increased phonon scattering can be achieved by introducing mass-fluctuation [13, 14], nanostructuring [15, 16], creating the hierarchical nano/meso-scale architectures [17] and intrinsic bond anharmonicity [18, 19]. However, the highest reported zT value is still below 3, which limits the usage of thermoelectric energy conversion to niche applications [20]. Low dimensional materials exhibit combination of both the enhancement in power factor due to quantum confinement effect, wherein flow of charge carriers is restricted and placed in a potential well with infinitely high walls, and the increased phonon scattering, led to significantly low thermal conductivity, by a large number of interfaces and grain boundaries [11, 21]. Confinement in the dimensionality significantly distorts the density of states of a given material and introduces the large interfaces that scatter phonon effectively rather than electrons and reduces the thermal conductivity drastically. For instance, Fig. 8.1a illustrates the schematic diagram of 3D (bulk), 2D (quantum well), 1D (quantum wire) and 0D (quantum dot) solid. Their corresponding DOS is given in Fig. 8.1b. In low dimensional systems, the thermoelectric parameters, such as S, σ and κ_{total} can be effectively tailored independently. Previous literature reports revealed that ultrathin nanosheets and nanofilms can exhibit enhanced densities of states (DOS) and carrier mobility (μ) compared with their bulk counterparts [22]. Significant increment in Seebeck coefficient has been achieved for n-type PbTe/Pb$_{0.927}$Eu$_{0.073}$Te multiple quantum wells (MQW) as compared to bulk PbTe material [10], which is consistent with theoretical calculations, as shown in Fig. 8.1c. A zT of 0.9 at 300 K and 2.0 at 550 K have been achieved in PbSe$_{0.98}$Te$_{0.02}$/PbTe quantum-dot structures [23, 24]. Further, the theoretical calculations suggested that Bi$_2$Te$_3$-based quantum well (2D) and quantum wire 1D materials would exhibit extremely high zT as compared to bulk materials due to quantum confinement induced remarkable enhancement in power factor and large interfaces driven low thermal conductivity. Figure 8.1d presents the exponential increase in zT of Bi$_2$Te$_3$ with decrease in thickness or diameter of quantum well (or quantum wire), respectively [9–11, 25, 26]. Thin film superlattices structures utilize the acoustic mismatch between the superlattice components to reduce κ_{lat} while retaining the electrical transport intact (phonon-blocking/electron-transmitting) [24]. These result in high thermoelectric performance which is rather impossible using the conventional alloying approach, thereby potentially eliminating alloy scattering of carriers [5].

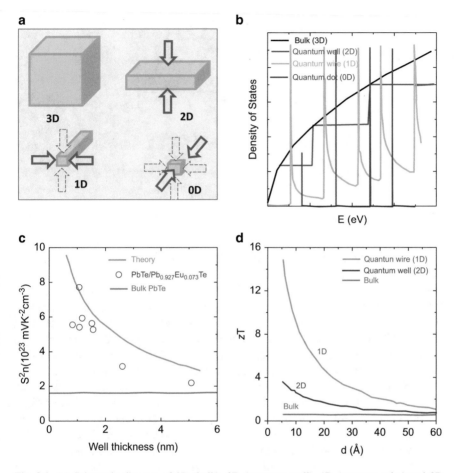

Fig. 8.1 (**a**) Schematic diagram of 3D (bulk), 2D (quantum well), 1D (quantum wire) and 0D (quantum dot) solid. (**b**) DOS vs. E plot for 3D, 2D, 1D and 0D, (**c**) the plot of S^2n vs. well thickness for n-type PbTe/Pb$_{0.927}$Eu$_{0.073}$Te multiple quantum wells (MQW) at 300 K, (**d**) calculated zT as a function of thickness/diameters of 3D, 2D and 1D for Bi$_2$Te$_3$-based materials [9–11]

8.2 Thermoelectric Transport Mechanism in Nanosheets

Ideas in using nanosheets (nanofilms) to progress the field of thermoelectrics through the enhancement of electrical conductivity and reduction of thermal conductivity were first discussed by Dresselhaus, Harman and Venkatasubramanian [4]. According to the Hicks–Dresselhaus model [11, 27], Seebeck coefficient for quantum-well structures increases linearly with decreasing the material thickness because of the quantum confinement effect, where S is directly proportional to the energy derivative of the density of states (DOS) at the Fermi level (E_F) (Mott–Jones relation, Eq. (8.1)) [11, 28].

$$S = \frac{\pi^2}{3} \frac{k_B}{q} k_B T \left\{ \frac{d\left[\ln(\sigma(E))\right]}{dE} \right\}_{E=E_F}$$ (8.1)

$$= \frac{\pi^2}{3} \frac{k_B}{q} k_B T \left\{ \frac{1}{n} \frac{dn(E)}{dE} + \frac{1}{\mu} \frac{d\mu(E)}{dE} \right\}_{E=E_F}$$

$$S = \frac{8\pi^2 k_B^2}{3eh^2} m_d^* T \left(\frac{\pi}{3n} \right)^{2/3}$$ (8.2)

$$m_d^* \alpha \left[\frac{g(E)}{\left(\sqrt{2E}\right)} \right]^{3/2}$$ (8.3)

Here, k_B is the Boltzmann constant, h is the Planck constant, m_d^* is the density of state effective mass of carriers, n is the carrier concentration and T is the absolute temperature. From Eq. (8.1), it is clear that S can be enhanced via the two mechanisms, (1) energy-dependence of carrier density, $n(E)$ and (2) energy-dependence of mobility, $\mu(E)$. Here $n(E)$ is associated with density of states, $g(E)$ $n(E) = g(E)f(E)$, where $g(E)$ is the DOS per unit volume and per unit energy; $f(E)$ is the Fermi function, whereas $\mu(E)$ depends on the energy of the charge carriers. As the best TE material falls under degenerate semiconductors, change in density of state, $g(E)$, has a direct consequence of the change in density of state effective mass (m_d^*) which significantly alters the Seebeck coefficient, as per Eqs. (8.2) and (8.3).

The enhancement of the Seebeck coefficient of nanosheet (nanofilm) sample originates from the steplike $n(E)$, where $g(E)$ can be improved via carrierpocket engineering. The anisotropic nature of the band structure offers a possibility of tuning the relative contributions of different carrier pockets near the Fermi surface by changing the several parameters, such as the growth direction. The sharp rise in the local DOS leads to an escalation in the Seebeck coefficient with minimal change in the carrier concentration, thus decoupling the electrical conductivity and the Seebeck coefficient [29, 30]. Hence, true atomic thickness would create a two-dimensional (2D) electron gas in the single layer, which would improve S for the ultrathin layered materials [22]. Furthermore, the potential barriers at the interfaces and boundaries of the nanofilm would be expected to filter out low-energy carriers (cold carriers) and transmit high-energy carriers (hot carriers), resulting in a further increased S [31]. This high carrier mobility derived from the electronic structure has a negligible impact on phonon transport. Therefore, the μ/κ_{lat} values of a nanosheets are always much higher than those of bulk materials [5]. Furthermore, the presence of the enormous grain boundaries and interfaces in the nanofilms scatters short/mid/long-wavelength phonons, thus contributing to a decrease in κ_{lat} with a minimal detrimental effect on the electrical properties [22]. Inorganic nanosheets usually exhibit surface structural disorder, allowing for efficient scattering of short-wavelength phonons, which decreases κ_{lat}.

Moreover, the thin film interfaces influence both in-plane and out-of-plane thermal conductivity [5]. Interface effects can be classified into three categories: specular, diffuse and hybrid (partially specular and partially diffuse). At a specular interface, phonons transport (transmitted or reflected) at the layer interface following Snell's law, when polarization changes are neglected [32]. When the incident angle is beyond a critical angle, the phonons should reflect totally and not cross the interface. At a diffuse interface, the thermal conductivity along the layers is irrelevant to the transmissivity and reflectivity. Therefore, the thermal transport properties can be estimated from the individual layers separately. However, at a hybrid interface, the energy flux of phonons between layers due to the transmissivity cannot be ignored. Here, the partial diffusion behaves as a barrier to weaken the continuity between layers. Nanosheets always possess partially specular and partially diffuse interfaces. Thus, the atomic-scale roughness is thought to be one of the main reasons for the reduction in the thermal conductivity of nanosheets materials. Considering that these mechanisms will not affect the carrier mobility, nanosheets have been regarded as one of the best materials to achieve the "phonon glass electron crystal" criterion [5].

8.3 Thin Films of Layered Chalcogenides Prepared by Physical Route

Bi_2Te_3-based systems have been well known for their superior TE properties even in the nanostructured bulk form [27]. Though many TE materials have been studied with considerably good TE performance, Bi_2Te_3-based alloys have been recognized as the state-of-the-art TE materials for room and/or low temperature applications [33]. In this part, we extensively discuss a unique structure of Bi_2Te_3 and the influence of physical deposition techniques and annealing temperature on TE properties of Bi_2Te_3-based thin films. The Bi_2Te_3 crystallizes in the rhombohedral crystal structure with the space group of R-3m with the lattice constant in the c-axis (30.48 Å) is much higher than that of the a, b-axis (4.38 Å). Presence of unique structural anisotropy in Bi_2Te_3 leads to large carrier mobility and electrical conductivity along ab-plane as compared to c-axis [29, 34]. Further, it has a hexagonal layered structure with stacking of quintuple layers of covalently bonded Te(1)–Bi–Te(2)–Bi–Te(1); each quintuple layer is connected by the weak van der Waals interactions [35].

The deposition of Bi_2Te_3-based films can be performed by various physical deposition techniques such as sputtering (magnetron/RF magnetron/co-sputtering/ion beam), evaporation (thermal/e-beam/flash), pulsed laser deposition (PLD) and molecular beam epitaxy (MBE) [36–49]. Among them, every technique has its own advantages and disadvantages. As is known, transport properties of thin films prepared by physical deposition techniques mostly depend on composition, crystal structure, substrate temperature, Te and Bi defects, annealing temperature, decomposition or evaporation rate of starting elements and even deposition techniques for some extent. In case of thermoelectric properties, thin film metrology also plays a key role as in-plane measurement of S, σ and κ_{total} is slightly easier than measurement in out of plane, where maintaining temperature difference is of great challenge. Thus, most of the research groups work on thin film-based thermoelectrics and report mostly the electronic-transport properties, such as S, σ and σS^2 due to challenge in the measurement of thermal conductivity. As TE properties strongly depend on the composition of materials, it is important to control an evaporation rate of Bi, Sb and Te during deposition of Bi_2Te_3 and $Bi_{2-x}Sb_xTe_3$-based thin films since Te, Bi and Sb possess different vapour pressures. Thus, reports showed that Te content in 60–65 at% gives an optimum composition of Bi_2Te_3, comparable with initial composition, with reasonably high zT [42–44].

Further thickness dependent thermoelectric properties of Bi_2Te_3-based thin films were also studied in detail as thick of the films will effectively control the transport properties. It has been further reported that co-evaporation method can be an efficient route to prepare Bi_2Te_3-based materials where evaporation rate of both Bi and Te can be simultaneously controlled during deposition, and nominal composition with optimum thickness could also be maintained. Deposition methods with the nominal composition of Bi_2Te_3-based materials and their corresponding thermoelectric properties are listed in Table 8.1. When compared to all other techniques co-evaporation and co-sputtering techniques seem to have better composition control and thereby notable high TE figure of merits, zTs of 0.91 and 0.81 for n-type Bi_2Te_3, 0.87 [48, 49, 57] for p-type $Bi_{0.5}Sb_{1.5}Te_3$. Nevertheless, p-type Bi_2Te_3/Sb_2Te_3 superlattices with individual layers of 10 Å prepared by low-temperature growth process have still being considered a predominant material with a record-high zT of $\sim$2.4 at 300 K [4].

Zhang et al. have successfully deposited the (00l) oriented Bi_2Te_3 thin films by magnetron co-sputtering method and compared with ordinary film and bulk target materials [51]. They have achieved the maximum power factor (σS^2) of 33.7 μW/cmK^2 at 360 K which is four times higher than the ordinary film and notably high compared to SPS sintered Bi_2Te_3 bulk material. This huge rise in σS^2 is due to the increase in mobility from $\sim$23 cm^2/V·s (ordinary film) to $\sim$52 cm^2/V·s for (00l) oriented Bi_2Te_3 thin film as unique anisotropy associated high mobility in ab-plane than c-axis. Further, the calculated cross-plane thermal conductivity was about 0.81 W/mK which is notably low as compared to p-type Bi_2Te_3/Sb_2Te_3 superlattices (0.95 W/mK) [4, 51]. The Bi_2Te_3 thin films with optimum 57 at% of Te content deposited on flexible polyimide substrate by vacuum thermal evaporation showed the σS^2 value of 12 μW/cmK^2 at 300 K. Moreover, the electrodeposited Bi_2Te_3-based thin films and nanowire arrays have shown the S value of -70 μV/K with σS^2 of 5.63 μW/cmK^2 [52]. The influence of deposition methods of prepared Bi_2Te_3-based thin films and the respective TE properties (σ, S and σS^2) have been illustrated in Fig. 8.2 and summarized in Table 8.2.

Table 8.1 Deposition methods, nominal composition and thermoelectric properties of n- and p-type Bi_2Te_3-based materials [50]

Composition	Method	Temp. (K)	σ $((10^4)$ S/m)	S (μV/K)	$S^2\sigma$ (μW/cm·K^2)
n-Bi_2Te_3 bulk target [51]	As melted	380	7.4	−130	10.8
n-Bi_2Te_3	Ordinary film	360	4	−140	7.2
n-(00l) oriented Bi_2Te_3	Magnetron co-sputtering	360	8.5	−230	33.7
n-$Bi_2(Te_{0.6}Se_{0.4})_3$ [36]	Vacuum thermal evaporation	300	700	200 (473 K)	35.3
n-Bi_2Te_3 [37]	RF sputtering	300	6.74	−118.6	9.5
n-Bi_2Te_3 [38]	e-beam evaporation	300	62.5	−19	2.3
n-Bi_2Te_3 [39]	RF magnetron sputtering	300	6.6	−38	1.1
		423	6.1	−80	4
		463	4.8	−80	3
		543	4.5	−160	13
n-Bi_2Te_3 [40]	Co-sputtering	303	3.2	−50	0.8
		423	3.2	−68	1.47
		573	3.5	−89	2.77
		673	5.5	−122	8.86
	Annealed at 673 K	423 (max)	7.4	−224	32.8
n-Bi_2Te_3 [52]	Electro-deposition	300	11.49	−70	5.63
n-Bi_2Te_3	RF magnetron sputtering	300	2.5	−150	5.62
n-Bi_2Te_3 [53]	Pulsed laser deposition	300	6.6	−194	25
p-$Bi_{0.3}Sb_{1.7}Te_3$		300	9.09	200	35.1
n-Te embedded Bi_2Te_3 [54]	Molecular beam epitaxy	300	6.5	205	27.3
		473	8.25	231	44.02
		523	7.1	190	25.63
n-Bi_2Te_3 [41]	RF magnetron sputtering	323	3.2	−94	27.1
p-$Bi_{1.5}Sb_{0.5}Te_3$ [55]	Thermal evaporation (45° tilt angle)	300	8.6	262	58.6
	(60° tilt angle)		7.9	238	45.2
	(90° tilt angle)		7.5	213	33.5
p-$(Bi/Sb)_2Te_3$ [56]	Thermal evaporation (45° tilt angle)	300	8.8	246	54.2
	(60° tilt angle)		7.4	221	36.5
	(90° tilt angle)		5.1	204	21.5

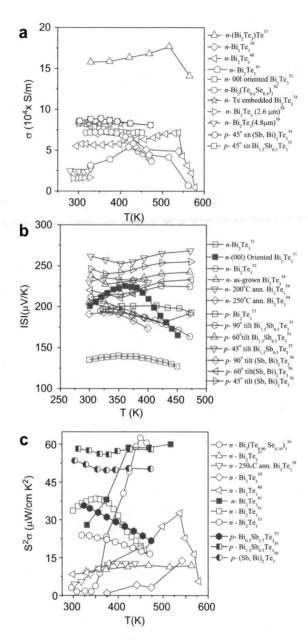

Fig. 8.2 Thermoelectric properties, (**a**) electrical conductivity (σ), (**b**) Seebeck coefficient (S) and (**c**) power factor ($S^2\sigma$) of Bi$_2$Te$_3$-based thin films deposited by various techniques

Table 8.2 Physical deposition methods, composition, applied temperature, annealing temperature and thermoelectric properties of Bi_2Te_3-based thin films [50]

Deposition methods	Composition	α ($\mu V/K$)	ρ ($\mu \Omega m$)	$S^2\sigma$ (10^{-3} W/m·K)	zT (300 K)
Co-evaporation	n-Bi_2Te_3	-228	28.3	1.8	–
Co-sputtering	n-Bi_2Te_3	-55	10	0.3	–
Co-evaporation	n-Bi_2Te_3	-220	10.6	4.57	0.91
Flash	n-$Bi_2Te_{2.72}Se_{0.3}$	-200	15	2.7	–
Sputtering	n-$Bi_{1.8}Sb_{0.2}Te_{2.7}Se_{0.3}$	-235	47	1.2	–
Co-sputtering	n-Bi_2Te_3	-160	16.3	1.6	–
Sputtering	n-$Bi_2Se_{0.3}Te_{2.7}$	-160	20	1.3	–
Co-evaporation	n-Bi_2Te_3	-228	13.0	4.0	0.81
Flash	p-$Bi_{0.5}Sb_{1.5}Te_3$	240	12	4.8	–
Co-sputtering	p-$(Bi/Sb)_2Te_3$	175	12.1	2.5	–
Flash	p-$Bi_{0.5}Sb_{1.5}Te_3$	230	17	3.1	0.87
Sputtering	p-$Bi_{0.5}Sb_{1.5}Te_3$	210	25	1.8	–

Pulsed laser deposited n-Bi_2Te_3 and p-$Bi_{0.3}Sb_{1.7}Te$ c-axis oriented thin films have shown significantly high TE performance [53]. For instance, in-plane measured σ, S and σS^2 of n-Bi_2Te_3 films are 6.6×10^4 S/m, -194 μV/K and 25.2 μW/cmK^2 [58], whereas p-$Bi_{0.3}Sb_{1.7}Te$ showed σ, S and σS^2 values of 9.09×10^4 S/m, 200 μV/K and 33.1 μW/cmK^2. Further, as annealing temperature after deposition, substrate temperature during deposition and thickness of films significantly influence the TE properties of prepared thin films, Hyejin Choi et al. have recently studied the effect of annealing temperature by controlling the each layer of Bi (3 Å) and Te (9 Å) on SiO_2/Si substrate using molecular beam epitaxial method on TE properties of Te embedded Bi_2Te_3 thin films [54]. The formed epitaxial Bi_2Te_3 thin film with heterojunctions has exhibited the maximum zT of $\sim$2.27 at 375 K due to significant phonon scattering rather than electron scattering, led to high σ and notably low κ_{total}. In particular, as-grown Bi_2Te_3 thin film showed σ, S, $S^2\sigma$ and κ_{total} values of 6.5×10^4 S/m, 205 μV/K, 27.3 μW/cmK^2 and $\sim$0.9 W/mK, whereas the film thus annealed at 473 K exhibited σ, S, $S^2\sigma$ and κ_{total} values of 8.25×10^4 S/m, 231 μV/K, 44.02 μW/cmK^2 and $\sim$0.75 W/mK. It is clear that Te embedded Bi_2Te_3 nano-grain structured epitaxial thin film showed the record-high TE performance due to increased phonon scattering rather than carrier scattering at grain boundaries and interfaces.

Further, it has been shown that grown angles of the array of nanowire or bundles of deposited Bi_2Te_3 thin films also alter the TE properties. Recently, Ming Tan et al. have reported that hierarchical p-$Bi_{1.5}Sb_{0.5}Te$ nanopillar array and p-$(Bi/Sb)_2Te_3$ nanowires array with well-oriented growth prepared by vacuum thermal evaporation method with a certain tilt angle to substrate showed extremely high zTs of 1.61 and 1.72 at 300 K [55, 56]. These observed huge enhancements in zT for tilt-structure of p-$Bi_{1.5}Sb_{0.5}Te$ array are due to preferred phonon scattering than carriers scattering in heterojunctions, and tilt-induced change in Fermi level enhances the flow of carriers and thereby σ increases drastically, which altogether supports to have high TE performance [59]. Presence of surface states at an energetic position above the conduction band edge in $(Sb,Bi)_2Te_3$ material exhibits a charge transfer from the surface state to the bulk [60]. So, electrical conductivity remarkably increases due to electron accumulation at the surface of nanowires with a high surface-to-volume ratio. This huge increase in σ from 5.1×10^4 S/m for a tilt angle of $90°$ to 9.1×10^4 S/m for a tilt-angle of $45°$ led to record-high zTs of 1.72 for c-axis oriented Bi_2Te_3 nanowire arrays (Fig. 8.2) [56].

8.4 Nanosheets of Layered Chalcogenides Synthesized by Chemical Route

Although thin-film thermoelectric materials offer tremendous scope for zT enhancement, questions remain on the precision of the zT reported due to experimental difficulties in measuring the properties correctly. Furthermore, most of the above methods require expensive apparatus and are not easy to control. Synthesis of large-area 2D metal chalcogenide thin films with controlled growth is a major challenge in this field of research. Synthesis based on the vapour processes such as molecular beam epitaxy has been successful in the preparation of large-area uniform thin films of 2D materials. However, this process requires synthesis conditions of high temperature, vacuum and specific substrates for materials growth. In addition, it is limited in the robustness of scale-up because the film growth takes a long time and the fine control of the film thickness over the entire large-size substrate is difficult. Additionally, materials fabricated by sophisticated techniques are not easily incorporated into commercial devices because they are slow and expensive to fabricate, and they cannot be fabricated in sufficient quantities. Solution phase synthesis has emerged as a promising alternative to the above-mentioned preparation technique. The solution-based process has several technological advantages: low temperature synthesis under atmospheric conditions, and more diversity in the material species to be synthesized. In the following section, we will be focusing on the thermoelectric properties of recently synthesized solution processed metal chalcogenides nanosheets.

8.4.1 Bismuth Telluride

Promising TE performance of bismuth telluride (Bi_2Te_3) thin film has motivated researchers to study TE properties of Bi_2Te_3 nanosheets. p- and n-type nanostructured bulk Bi_2Te_3-based materials with zT of 1.1 have been realized by bottom-up assembly of rapidly synthesized single-crystalline nanoplates of sulphur-doped

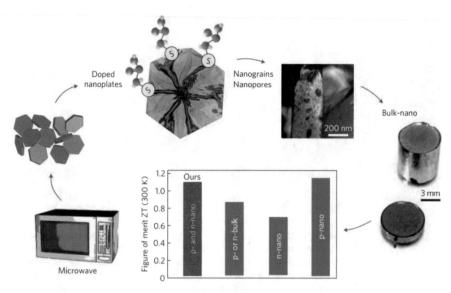

Fig. 8.3 Schematic demonstration of the scalable synthesis technique used to obtain both n- and p-type bulk thermoelectric nanomaterials $Bi_2Te_{3-x}S_x$ with high figures of merit. Microwave synthesis of $Bi_2Te_{3-x}S_x$ nanoplates followed by cold-pressing and sintering yields dense bulk pellets (density is $92 \pm 3\%$ of theoretical density) with nanostructured grains. Sulphur doping in Bi_2Te_3 optimizes the electrical conductivity, Seebeck coefficient and majority carrier type, while nanostructuring results in very low κ_{total}. The graph compares the best ZT ($\sim zT$) of p- and n-type nanomaterials with those of the best p- and n-bulk materials, denoted as p- or n-bulk, nanoparticle-dispersed n-bulk, referred to as n-nano, and a p-type ball-milled alloy, denoted as p-nano. Adapted with permission from Ref. [61] © 2012, Nature Publishing Group

Bi/Sb telluride [61]. Bismuth chloride ($BiCl_3$) and antimony chloride ($SbCl_3$) have been used as metal precursors in inexpensive organic solvents for rapid preparation of large amount of sulphur-doped pnictogen chalcogenide (V_2VI_3) nanosheets to prepare single- and multi-component nanostructured bulk TE materials (Fig. 8.3). Microwave stimulation enhances the rate of reaction between molecularly ligated chalcogen and pnictogen complexes with thioglycolic acid (TGA) in the presence of a high-boiling solvent. Nanostructured Bi_2Te_3 pellets showed lower lattice thermal conductivity and high power factors, comparable to its bulk counterpart. Nanostructuring and sulphur doping result in these remarkable charge-carrier-crystal and phonon-glass behaviours. Thus, Bi_2Te_3 nanosheets exhibit 250% higher zT than their non-nanostructured bulk counterparts and state-of-the-art alloys (Fig. 8.3).

Spark plasma sintering (SPS) is well known to be a very useful technique for preparation of nanostructured bulk TE materials because of its fast heating and cooling rates, which allow rapid sintering, avoid undesirable grain growth emerging from a long sintering process at high temperatures. The grain growth can be manipulated by varying the SPS conditions, allowing to study the effect of grain size and density on the TE properties of nanostructured bulk materials. Recently

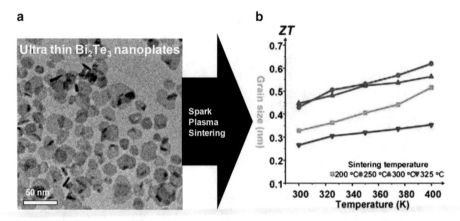

Fig. 8.4 Large-scale synthesis of ultrathin Bi_2Te_3 nanoplates and subsequent spark plasma sintering to fabricate *n*-type nanostructured bulk thermoelectric materials. (**a**) The transmission electron microscopy (TEM) image of as-synthesized ultrathin Bi_2Te_3 nanoplates. (**b**) Temperature dependence of *ZT* (*zT*) of nanostructured bulk Bi_2Te_3 sintered at 200 °C (green square), 250 °C (red circle), 300 °C (blue upward triangle) and 325 °C (purple downward triangle). Adapted with permission from Ref. [31] © 2012, American Chemical Society

Son et al. have reported TE properties of SPS-processed ultrathin (1–3 nm) *n*-type Bi_2Te_3 nanoplates (Fig. 8.4) [31]. Bi_2Te_3 nanoplates have been synthesized by the reaction between bismuth dodecanethiolate and tri-n-octylphosphine telluride in the presence of oleylamine. The highest *zT* of 0.62 has been achieved in the SPS-processed sample sintered at 250 °C, which is one of the highest values among those reported for *n*-type chemically synthesized nanosheets (Fig. 8.4).

8.4.2 Bismuth Selenide

Bismuth selenide (Bi_2Se_3), another narrow band gap semiconductor (~0.3 eV), has potential application in the field of thermoelectrics [36–39]. Like Bi_2Te_3, it is also a 3D topological insulator (TI), where metallic surface states are protected by the time-reversal symmetry [22]. Bi_2Se_3 is a layered anisotropic material (space group: R-3 m) accommodating quintuple layers (QL) each of which are having thickness of ~1 nm and composed of five covalently bonded atomic planes [Se2–Bi–Se1–Bi–Se2]. Thus, synthesis of atomically thin Bi_2Se_3 nanosheets is desirable for TE application. Sun et al. synthesized single layered Bi_2Se_3 via a scalable intercalation/exfoliation strategy, by using Li-intercalated Bi_2Se_3 microplates as an intermediate precursor [22]. Bi_2Se_3 nanosheets sample shows high carrier mobility (μ) of ~6000 cm^2 V^{-1} s^{-1} which can be attributed to the 2D electron gas like nature [22, 62]. The single-layer-based (SLB) Bi_2Se_3 composite has shown a huge improvement in electric transport properties and ultralow thermal conductivity than that of the bulk counterpart over the entire temperature range. Thus, the σ/κ_{total} ratio for the SLB composite is superior to that of the bulk material over the whole temperature range. $|S|$ for the SLB composite gradually increases from 90 μV/K at 300 K to 121 μV/K at 400 K, which is greater than that of bulk material ($\sim$98.5 μV/K at 400 K). Thus, zT for the SLB composite is higher than that for the bulk material over the entire temperature range. Specifically, the SLB composite exhibits zT of 0.35 at 400 K, which is eight times larger than zT of the bulk Bi_2Se_3 and higher than the previously reported values for pure Bi_2Se_3 nanostructures, suggesting the advantage of the atomically thick SLB structure over bulk one (Fig. 8.5).

Jana et al. have investigated the TE properties of few-layer Bi_2Se_3 nanosheets, synthesized via a green ionothermal reaction in the water-soluble, room-temperature ionic liquid, 1-ethyl-3-methylimidazolium tetrafluoroborate ([EMIM][BF$_4$]) [62]. Ionothermal reaction of bismuth acetate and selenourea in [EMIM][BF$_4$] results in ultrathin few-layer (3–5 layer) Bi_2Se_3 nanosheets. The high σ of Bi_2Se_3 nanosheets originates from the presence of metallic surface states that offer high mobility and

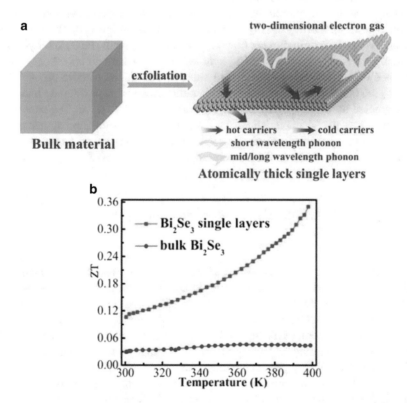

Fig. 8.5 (a) Schematic representation of exfoliation technique and electronic and phonon transport in single layer Bi_2Se_3. (b) Temperature dependent ZT (zT) of bulk and single layered Bi_2Se_3. Adapted with permission from Ref. [22] © 2012, American Chemical Society

scattering resistant carriers transport. In addition to this, surface defects, nanoscale grain boundaries and interfaces effectively scatter the heat carrying phonons, thereby decreasing κ_{total} to $\sim$0.4 Wm^{-1} K^{-1} near room temperature.

8.4.3 Antimony Telluride

Antimony telluride (Sb$_2$Te$_3$) is a narrow-band gap ($\sim$0.28 eV) layered semiconductor with tetradymite structure [63]. Single-crystalline Sb$_2$Te$_3$ nanosheets with lateral dimension of 300–500 nm and thicknesses of 50–70 nm were rapidly synthesized by a microwave-assisted reaction of SbCl$_3$, Na$_2$TeO$_3$ and N$_2$H$_4$.5H$_2$O in ethylene glycol at 200 °C [64]. For thermoelectric measurements, nanostructured bulk Sb$_2$Te$_3$ was prepared from Sb$_2$Te$_3$ nanosheets via SPS. High electrical conductivity ($\sim$2.49 $\times$ 10^4 Sm^{-1}), high Seebeck coefficient ($\sim$210 μVK^{-1}) and low thermal conductivity ($\sim$0.76 Wm^{-1} K^{-1}) at 420 K were achieved, which resulted in zT of 0.58 at 420 K for the Sb$_2$Te$_3$ nanosheets sample.

Dong et al. have explored rapid microwave-assisted solvothermal synthesis and thermoelectric properties of Sb$_2$Te$_3$ nanosheets [65]. Antimony trichloride anhydrous (SbCl$_3$), tellurium dioxide (TeO$_2$), polyethylene glycol (PEG) and hydrazine hydrate (N$_2$H$_4$.5H$_2$O) were used to synthesize the nanostructured Sb$_2$Te$_3$ in the presence of ethylene glycol (EG) as a solvent. The thermoelectric properties of the Sb$_2$Te$_3$ pellet obtained by room-temperature pressing of powdered nanosheets were studied at the temperatures ranging from 300 to 450 K. Sb$_2$Te$_3$ nanosheets samples show the characteristics of p-type semiconductors. Sb$_2$Te$_3$ nanosheets exhibit Seebeck coefficient of $\sim$194–245 μVK^{-1} and power factor of 0.48–1.14 $\times$ 10^{-4} Wm^{-1} K^{-2} in the temperature range of 300–450 K.

8.4.4 Solid Solutions and Nanocomposites of Bi$_2$Te$_3$ and Bi$_2$Se$_3$

Solid solutions and nanocomposites of Bi$_2$Te$_3$, Bi$_2$Se$_3$ and Sb$_2$Te$_3$ nanosheets have been explored for thermoelectric application. Because of the isomorphic crystal structure of Bi$_2$Te$_3$ and Bi$_2$Se$_3$, the solubility of Se in Bi$_2$Te$_3$ results in a modification of the crystal lattice and electronic density of states (DOS), which is beneficial for reduction of the bipolar conduction and is an important factor for the advancement of thermoelectric properties of Bi$_2$Te$_3$ alloys. Min et al. have presented TE properties of Bi$_2$Te$_{3-x}$Se$_x$ nanocomposite pellets [66]. Figure 8.6 presents TE properties of the as-prepared pellets by mixing appropriate ratios of the Bi$_2$Te$_3$ and Bi$_2$Se$_3$ nanoflakes. The electrical conductivities (σ) of the nanocomposites systematically increased in the range of 200–440 Scm^{-1} at room temperature as the Bi$_2$Se$_3$ amount has been enhanced (Fig. 8.6a). The Seebeck coefficient is negative

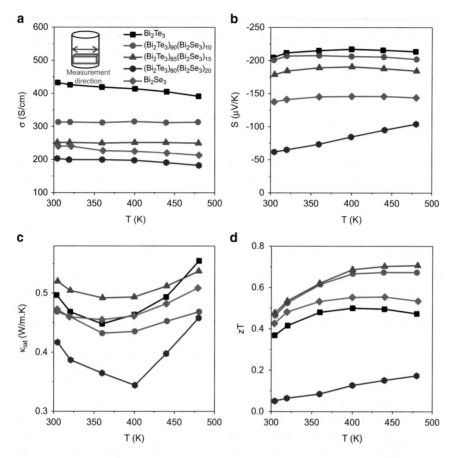

Fig. 8.6 Temperature dependent (**a**) electrical conductivities (σ), (**b**) Seebeck coefficients (S), (**c**) lattice thermal conductivities (κ_{lat}) and (**d**) thermoelectric figure of merit (zT) for nanocomposite of Bi_2Te_3 and Bi_2Se_3. Inset in (**a**) shows schematic of sintered pellet along with thermoelectric measurement directions. Adapted with permission from Ref. [66] © 2013, John Wiley and Sons

(Fig. 8.6b) indicating electrons as the major charge carriers. The absolute value of S increased as the measurement temperature has been raised to 360 K. S has increased with the increasing composition of the Bi_2Se_3 nanoflakes. The maximum value of S was found for $(Bi_2Te_3)_{85}(Bi_2Se_3)_{15}$ which can be attributed to the carrier energy filtering effect originating from the presence of Bi_2Se_3 nanoflakes in the Bi_2Te_3 matrix. The maximum power factor (1.2 mW m^{-1} K^{-2} at 400 K) in this study was obtained in $(Bi_2Te_3)_{85}(Bi_2Se_3)_{15}$. The κ_{total} values of the nanocomposites ranged from 0.55 to 0.68 Wm^{-1} K^{-1} at room temperature, which is significantly lower than that of the bulk Bi_2Te_3 with micro-sized grains ($\sim$1.5 Wm^{-1} K^{-1}). This reduction of κ_{total} is attributed to the reduction of κ_{lat} due to the phonon scattering at the interfaces of the randomly oriented nanograins (5–20 nm). Room temperature

κ_{lat} values of the nanocomposites decreased by about 40% (0.43–0.52 Wm^{-1} K^{-1}) (Fig. 8.6c). The reduced κ_{total} and the enhanced power factor result in maximum thermoelectric figure of merit (zT) of $\sim$0.71 at 480 K for $(Bi_2Te_3)_{90}(Bi_2Se_3)_{10}$ samples (Fig. 8.6d).

8.4.5 SnSe

SnSe, an environment-friendly layered chalcogenide, has drawn enormous consideration of the TE community with their high thermoelectric performance in single crystals [67–69]. SnSe crystallizes in layered orthorhombic crystal structure (space group *Pnma*, lattice parameters, $a = 11.502$ Å, $b = 4.15$ Å, $c = 4.45$ Å) at room temperature with sterically accommodated lone pair [69]. When temperature increases ($\sim$800 K), SnSe undergoes a second-order displacive phase transition to a higher symmetric five-fold coordinated *Cmcm* phase [69]. The two-atom-thick SnSe slabs are folded up and create a zigzag accordion-like projection along the crystallographic *b* axis. In both the structures these identical layers are weakly bound via weak van der Waals interaction and resulting in an anisotropic layered structure. Han et al. have recently developed a surfactant-free simple solution-based technique, using water as a solvent, for the synthesis of phase pure gram scale orthorhombic SnSe nanoplates [70]. Hot-pressed SnSe nanosheets exhibit outstanding electrical conductivity (σ) and power factors (σS^2). Recently, Chandra et al. have reported the solution phase synthesis and thermoelectric transport properties of two-dimensional (2D) ultrathin few-layer nanosheets (2–4 layers) of *n*-type SnSe using SnCl$_4$·5H$_2$O and SeO$_2$ as a preliminary precursor [71]. The *n*-type nature of the SnSe nanosheets arises from in situ chlorination during the synthesis. The carrier concentration of *n*-type SnSe has been significantly increased from 3.08×10^{17} cm^{-3} to 1.97×10^{18} cm^{-3} via Bi-doping which results in the significant rise of electrical conductivity and power factor (Fig. 8.7). Typically, $Sn_{0.94}Bi_{0.06}Se$ nanosheets ($\perp$ to pressing direction) have S value of -219 μV/K at 300 K, which increases to -285 μV/K at 719 K. Furthermore, Bi-doped nanosheets exhibit ultralow lattice thermal conductivity ($\sim$0.3 W/mK) throughout the temperature range of 300–720 K originating from the effective phonon scattering by interface of SnSe layers, nanoscale grain boundaries and point defects.

8.4.6 SnSe$_2$

Tin diselenide (SnSe$_2$) is an additional layered compound from the phase diagram of Sn-Se [72–74]. Our group has reported the synthesis of *n*-type ultrathin few-layer SnSe$_2$ nanosheets via the simple solution-based low temperature synthesis, which exhibits semiconducting electronic transport (Fig. 8.8) [72]. An ultralow

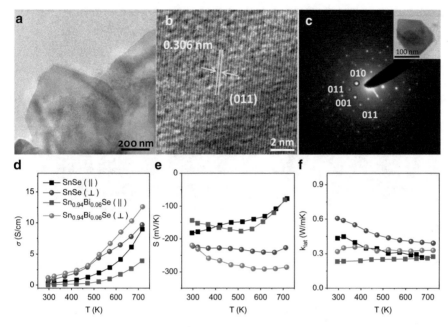

Fig. 8.7 (**a**) TEM image of $Sn_{0.94}Bi_{0.06}Se$ nanosheets. (**b**) HRTEM image of $Sn_{0.94}Bi_{0.06}Se$ nanosheets. (**c**) SAED pattern of a single $Sn_{0.94}Bi_{0.06}Se$ nanosheet. Inset of (**c**) shows the STEM image of single nanosheet of $Sn_{0.94}Bi_{0.06}Se$. Temperature dependent (**d**) electrical conductivity (σ), (**e**) Seebeck coefficient, (S) and (**f**) lattice thermal conductivity (κ_{lat}) of SnSe and $Sn_{0.94}Bi_{0.06}Se$ nanosheets measured along the SPS direction (indicated by squares) and perpendicular to the SPS direction (indicated by spheres). Adapted with permission from Ref. [71] © 2018, American Chemical Society

thermal conductivity (~ 0.67 Wm^{-1} K^{-1}) has been achieved due to the anisotropic layered structure, which results in effective phonon scattering at layered interface and grain boundaries. However, the low carrier concentration ($\approx 10^{18}$ cm^{-3}) and randomly oriented grains still result in a low power factor (≈ 150 μWm^{-1} K^{-2}). Thermoelectric properties of $SnSe_2$ have been improved by Kanatzidis and co-worker by simultaneously introducing a selenium (Se) deficiency and chlorine (Cl) doping in $SnSe_2$ nanoplate-based pellets, in which the nanoplates show a preferred orientation of the (001) planes along the primary surface of the pellet (in-plane) [73]. This yields a sharp increase in the in-plane electrical conductivity and power factor. The $SnSe_2$ nanoplate-based pellets have been prepared via a two-step process: (1) the synthesis of precursor bulk ingots by a vacuum-sealed high temperature melting process, followed by ball-milling grinding of the ingots; (2) obtaining the nanoplate-based pellets through the densification of ground precursors in SPS process. The electrical resistivity (ρ) of the Se-deficient and halogen-doped samples is much lower than that of the pristine $SnSe_2$ over the entire temperature range (Fig. 8.9). As a result, an improved in-plane thermoelectric figure of merit, zT_{max}, of 0.63 is

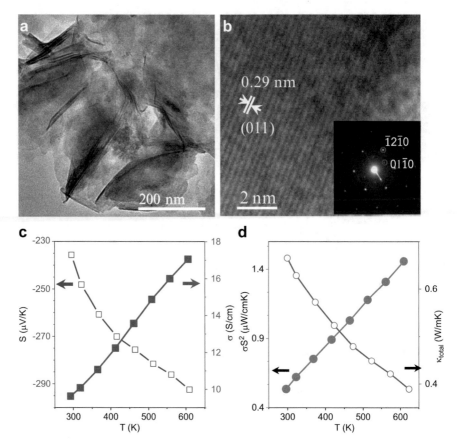

Fig. 8.8 (**a**) TEM image of Cl-doped SnSe$_2$ nanosheets. (**b**) HRTEM image of Cl-doped SnSe$_2$ nanosheets showing distance between (011) planes. Inset of (**b**) shows SAED pattern of a single nanosheet. Temperature dependent (**c**) Seebeck coefficient (S) and electrical conductivity (σ), and (**d**) power factor (σS^2) and thermal conductivity (κ_{total}) of SnSe$_2$ nanosheets. Adapted with permission from Ref. [72] © 2016, John Wiley and Sons

obtained for a 1.5 at% Cl-doped SnSe$_{1.95}$ pellet at 673 K, which is significantly higher than the corresponding in-plane zT of pure SnSe$_2$ (0.08) (Fig. 8.9).

8.4.7 Layered Intergrowth Chalcogenides

Layered intergrowth inorganic compounds in the homologous series have drawn the attention of thermoelectric community due to their ultralow thermal conductivity [75–79]. Phase homology is a collective demonstration of various structures built on the same structural principle with a certain module(s) expanding in different

Fig. 8.9 Temperature dependent (**a**) electrical resistivity (ρ) and (**b**) *ZT* (*zT*) of the 1.5 at% Cl doped SnSe$_{1.95}$ sample and *p*-type polycrystalline SnSe sample measured along in-plane and cross-plane direction. Adapted with permission from Ref. [73] © 2017, John Wiley and Sons

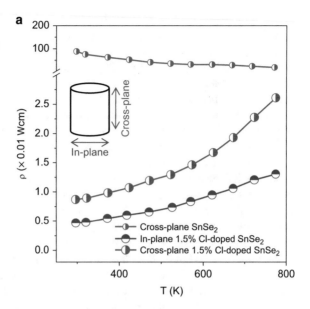

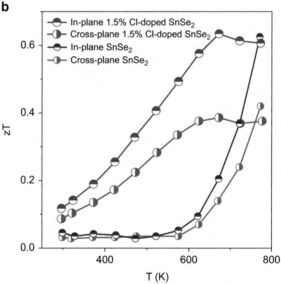

dimensions in standard increments. The modules are usually infinite rods, cluster blocks or layers, which can be consolidated in various ways to frame each member via coordination chemistry principles or structure building operators, reflection twinning, glide reflection twinning, cyclic twinning, unit cell intergrowth, etc. [75] The homologous series is represented by a mathematical formula which is able to produce each member [75, 80]. In solid-state inorganic chemistry, the use of phase homologies is a well-established approach to predict new compounds with a predictable structure. The general representation of some of the series includes $A_m(M'_{1+l}Se_{2+l})_{2m}(M''_{2l+n}Se_{2+3l+n})$ (A = alkali metal; M', M'' = main group element), $Cs_4(Bi_{2n+4}Te_{3n+6})$ and $(Sb_2Te_3)_m(Sb_2)_n$ [75, 80, 81]. These materials are structurally as well as electronically anisotropic in nature, which can have a positive impact on the thermoelectric properties.

Layered intergrowth compounds from the homologous series of quasi-binary $(A^{IV}B^{VI})_m(A^V{}_2B^{VI}{}_3)_n$ systems, where A^{IV} = Ge, Sn, Pb; A^V = Sb, Bi and B^{VI} = Se, Te, are predicted to show promising thermoelectric performance [75, 82]. These compounds crystallize in anisotropic layered tetradymite Bi_2Te_2S-type structures [82–86]. Most of the layered compounds in the above homologous series resemble natural van der Waals heterostructure and are anticipated to be 3D-topological insulators [87–90]. These compounds are predicted to show low κ_{lat} due to complex crystal structure and strong phonon scattering at the interfaces between the layers [82, 84]. However, it is really difficult to synthesize these compounds in pure phases through high temperature solid-state melting technique due to their incongruent melting nature (Fig. 8.10a) [83, 84]. Preparation of these compounds in nanosheets form can provide excess metallic surfaces with scattering resistant transport and high carrier mobility. Potential applications of these intergrowth nanosheets in the field of topological materials and thermoelectrics have grown interest to study the layered complex compounds in the pseudo-binary homologous series. As an example, we can consider $(SnTe)_m-(Bi_2Te_3)_n$ series, where $SnBi_2Te_4$, $SnBi_4Te_7$ and $SnBi_6Te_{10}$ are distinguished members (Fig. 8.10) [84]. They can also be visualized as intergrowths of SnTe-type rocksalt and Bi_2Te_3-type hexagonal phases, which are indeed natural heterostructures [84]. In the unit cell of $SnBi_2Te_4$ (rhombohedral structure, R-3m space group), septuple layers are stacked along the c-axis by van der Waals interactions. Each septuple layer is composed of seven covalently bonded atomic planes [Te2–Bi–Te1–Sn–Te1–Bi–Te2] with 1.13 nm thickness (Fig. 8.10b). Layered intergrowth compound, $SnBi_4Te_7$ [i.e. $(SnTe)_1(Bi_2Te_3)_2$] crystallizes in trigonal structure (P-3m1 space group) with a long unit cell (c = 2.406 nm). The $SnBi_4Te_7$ crystal structure contains two subunits, a quintuple layered (QL) Bi_2Te_3 and a septuple layered (SL) $SnBi_2Te_4$, assembled along the c-axis with the 5757 sequences by van der Waals interactions (Fig. 8.10c). In $SnBi_6Te_{10}$ [i.e. $(SnTe)_1(Bi_2Te_3)_3$], $SnBi_2Te_4$ and Bi_2Te_3 subunits are assembled along the c-axis in 7557 sequence (Fig. 8.10d). Similarly, several layered intergrowth compounds exist in the $(PbTe)_m-(Bi_2Te_3)_n$ and $(PbTe)_m-(Bi_2Se_3)_n$. Recently, 2D ultrathin nanosheets of various layered compounds from homologous $A_mBi_{2n}Te_{3n+m}$ have been synthesized by low temperature solution-based bottom up method and their thermoelectric properties have been investigated [83, 84]. The nanosheets were characterized using various experimental techniques (Fig. 8.11). Atomic force microscopy (AFM) and transmission electron microscopy (TEM) measurement indicate the ultrathin nature of the nanosheets. HAADF-STEM imaging confirms van der Waals heterostructured nature of nanosheets. Few-layer nanosheets indeed exhibit semiconducting electronic-transport properties with high carrier mobility (Fig. 8.12a, b). Long periodic intergrowth structure and effective phonon scattering

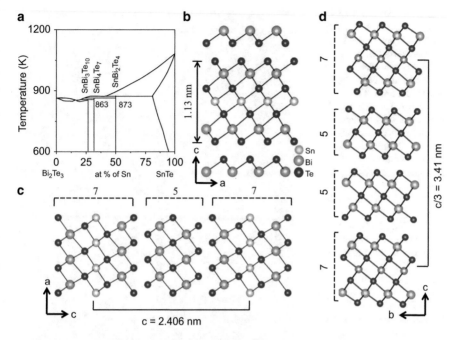

Fig. 8.10 (**a**) Existence of incongruently melting layered ternary intergrowth tin chalcogenide compounds in the SnTe–Bi$_2$Te$_3$ pseudo-binary phase diagram. (**b**) Crystal structure of SnBi$_2$Te$_4$ showing the 1.13 nm thick septuple atomic layers. (**c, d**) Crystal structure of natural van der Waals heterostructure SnBi$_4$Te$_7$ and SnBi$_6$Te$_{10}$ showing different stacking sequences of quintuple layers of Bi$_2$Te$_3$ and septuple layers of SnBi$_2$Te$_4$

at the interface lead to low thermal conductivity (κ_{lat} of 0.3–0.5 Wm^{-1} K^{-1}) for these 2D nanosheets (Fig. 8.12c).

8.4.8 BiCuSeO

Layered quaternary metal oxychalcogenide, MCuXO (M = La, Ce, Nd, Pr, Bi and In; X = Te, Se and S) is reported to show superior properties in various fields like thermoelectrics and superconductivity [91, 92]. BiCuSeO, a multiband semiconductor and a promising TE material from this oxychalcogenide family [93]. It crystallizes in ZrCuSiAs-type layered structure with a tetragonal unit cell (space group P4/nmm). BiCuSeO crystal structure comprises of insulating (Bi$_2$O$_2$)$^{2+}$ and conducting (Cu$_2$Se$_2$)$^{2-}$ layers, which are stacked along the crystallographic c-axis of the tetragonal cell (Fig. 8.13a) [93]. The (Bi$_2$O$_2$)$^{2+}$ layers are composed by distorted Bi$_4$O tetrahedra (fluorite type structure), whereas (Cu$_2$Se$_2$)$^{2-}$ layers contain distorted CuSe$_4$ tetrahedra (anti-fluorite type structure). The ultralow κ_{lat} originates from the scattering of phonons from the interface due to the layered structure, soft

Fig. 8.11 (a) AFM image of a SnBi$_2$Te$_4$ nanosheet. The inset in (a) shows the Tyndall light scattering effect of SnBi$_2$Te$_4$ nanosheet dispersed in toluene. (b) TEM image of a SnBi$_2$Te$_4$ nanosheet. The inset in (b) shows the SAED pattern of a single SnBi$_2$Te$_4$ nanosheet. (c) HRTEM image showing the crystalline nature of as-synthesized of SnBi$_2$Te$_4$ nanosheet. (d) HAADF-STEM image shows long-range ordered sequence of SnBi$_2$Te$_4$ blocks with a schematic of the atomic layer sequence in structures built from SnBi$_2$Te$_4$. (e) AFM image of a SnBi$_4$Te$_7$ nanosheet. (f) TEM image of SnBi$_4$Te$_7$ nanosheet. The inset in (f) shows the SAED pattern of a single SnBi$_4$Te$_7$ nanosheet. (g) HRTEM image of a SnBi$_4$Te$_7$ nanosheet. (h) HAADF-STEM image of SnBi$_4$Te$_7$ showing presence of 57 stacking of Bi$_2$Te$_3$ and SnBi$_2$Te$_4$. Adapted with permission from Ref. [84] © 2017, John Wiley and Sons

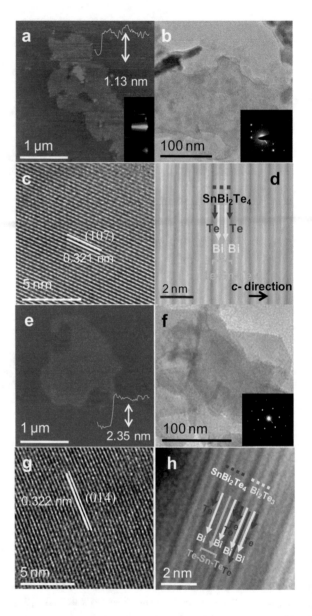

bonding and lattice anharmonicity, which result in low κ_{lat} in BiCuSeO. A recent study by Samanta et al. has explored the synthesis of few-layered ultrathin BiCuSeO nanosheets via a facile surfactant-free low temperature solvothermal synthesis [94]. Large-scale few-layered BiCuSeO nanosheets (Fig. 8.13b–d) have been synthesized by the reaction of Bi(NO$_3$)$_2$·5H$_2$O, Cu(NO$_3$)$_2$·3H$_2$O and selenourea in the presence of KOH/NaOH under solvothermal conditions. BiCuSeO nanosheets exhibit lower lattice thermal conductivity (0.55–0.4 Wm^{-1} K^{-1}) compared to that of the bulk

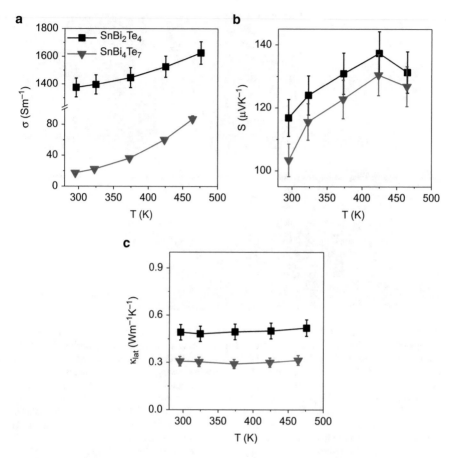

Fig. 8.12 Temperature-dependent (**a**) electrical conductivity (σ), (**b**) Seebeck coefficient (S) and (**c**) lattice thermal conductivity (κ_{lat}) of SnBi$_2$Te$_4$ and SnBi$_4$Te$_7$ nanosheets with 5% error bar. Adapted with permission from Ref. [84] © 2017, John Wiley and Sons

sample (Fig. 8.13f). Significant phonon scattering from the interfaces of the layers, bond anharmonicity and nanoscale grain boundaries resulted in low thermal conductivity in BiCuSeO nanosheets. Thus, nanosheets of pristine BiCuSeO demonstrate potential application in thermoelectric energy harvesting.

8.4.9   Cu$_2$Se

As one of the promising thermoelectric materials, Cu$_2$Se provides opportunities to overcome the global energy crisis via the conversion of waste heat into electricity [95]. Low temperature α-Cu$_2$Se phase has a complex monoclinic crystal structure

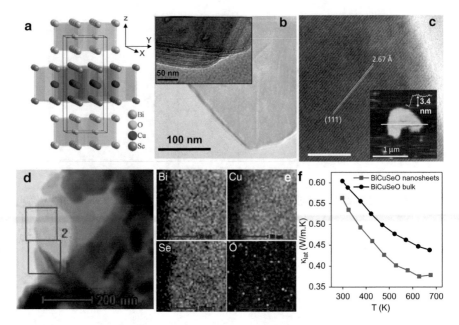

Fig. 8.13 (a) Crystal structure of BiCuSeO demonstrating the layered structure. (b) Ultrathin van der Waals heterostructured nanosheets of layered quaternary metal oxychalcogenide, BiCuSeO. Inset image shows that the bent edges of several BiCuSeO nanosheets are stacked one over another. (c) HRTEM image of the BiCuSeO nanosheets showing the crystalline nature and lattice spacing between (111) planes. The inset image demonstrates an AFM image of the BiCuSeO nanosheets and the height profile showing a thickness of ∼3.4 nm. (d) STEM image of the BiCuSeO nanosheets. (e) EDAX colour mapping for Bi, Cu, Se and O in the BiCuSeO nanosheets during STEM imaging (from the highlighted portion (numbered as 2) of the STEM image). (f) Temperature dependent lattice thermal conductivity (κ_{lat}) of BiCuSeO nanosheets (marked in red) and bulk BiCuSeO (marked in black). Adapted with permission from Ref. [94] © 2017, Royal Society of Chemistry

with 144 atoms per unit cell. When the temperature increases to 400 K, α-Cu_2Se transforms to a high temperature β-phase with Fm-3m space group. During the phase transition, Cu^+ ions assemble in the ordered stack along the <111> directions to form a simple anti-fluorite structure. Such a phase transformation is reversible through cooling or heating processes. In case of β-Cu_2Se crystal structure, Se atoms form a face-centred-cubic (FCC) frame and Cu^+ ions behave like liquid (high mobility) with a reduced phonon mean free path [96], which results in a low κ_{lat} value of 0.4–0.6 Wm^{-1} K^{-1}. Yang et al. have studied thermoelectric properties of SPS-processed β-phase Cu_2Se nanoplates [95]. Significantly enhanced phonon scattering achieved by high-density small-angle grain boundaries and dislocations efficiently block phonons with long and intermediate mean free paths. Moreover, Cu^+ ions strongly scatter phonons with a short mean free path. This full-spectrum phonon scattering has a negligible control over the electrical transport because electrons are having very short mean free path which can transport through grains.

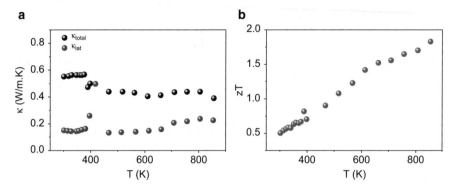

Fig. 8.14 Temperature dependent (**a**) thermal conductivities and (**b**) zT of Cu_2Se. Adapted with permission from Ref. [95] © 2015, Elsevier

Thus, an ultra-low κ_{lat} ($\sim$0.2 $Wm^{-1}\ K^{-1}$) and zT of 1.82 has been achieved in SPS-processed Cu_2Se pellets (Fig. 8.14). Such an enhanced zT could be attributed to its very low κ_{lat}, which benefits from the strong phonon scattering by high densities of small-angle grain boundaries and dislocations within the boundaries via nanostructure engineering.

8.5   Conclusions and Future Directions

In this chapter, we have discussed the key concepts, latest development and under-standing of the thermoelectric properties of metal chalcogenides nanosheets and 2D thin films prepared by physical and chemical routes. However, thermoelectric thin film metrology is yet to be reinvestigated as uncertainty in thin film measurement is still a challenging task, especially measurement of cross-plane TE properties of 2D materials. Nevertheless, Bi_2Te_3-based nanosheets and thin films are still considered as leading thermoelectric materials for near room temperature power generation and refrigeration applications, while $SnSe$, Cu_2Se nanosheets are observed to be potential candidates for the mid-to-high temperature power generation.

Generally, layered metal chalcogenides show intrinsically low thermal conductivity due to anisotropic structure and strong lattice anharmonicity. Intergrowth homologous chalcogenides are new candidates for thermoelectric applications as they exhibit low thermal conductivity due to long periodic intergrowth structure. The studies on the thermoelectric properties of layered chalcogenides are vibrant and open area of research. Although the thermal conductivity is intrinsically low for layered chalcogenides, attention should be given on improving the Seebeck coefficient by various innovative approaches such as electronic band valley conver-gence and exploration of resonance level in the electron structure. Many new layered materials with promising thermoelectric properties have been discovered, although

A. Banik et al.

a lot of work and progress need to be done for their practical realization in the form of thermoelectric module. In this context, the combined effort from chemistry, physics, materials scientists and engineers will certainly lead to the achievement of high performance 2D chalcogenide-based TE materials and fabrication of highly efficient TE devices for localized power generation, small-scale electronic systems and for booting-up the conventional combustion gasoline heat engines.

Acknowledgements The authors thank Nanomission, DST (SR/NM/TP-25/2016) and Sheikh Saqr Laboratory for financial support. A. B. thanks INSPIRE Programme for fellowship.

References

1. G. Tan, L.-D. Zhao, M.G. Kanatzidis, Chem. Rev. **116**(19), 12123 (2016)
2. C. Xiao, Z. Li, K. Li, P. Huang, Y. Xie, Acc. Chem. Res. **47**(4), 1287 (2014)
3. J. He, T.M. Tritt, Science **357**(6358), eaak9997 (2017)
4. R. Venkatasubramanian, E. Siivola, T. Colpitts, B. O'Quinn, Nature **413**(6856), 597 (2001)
5. Y. Zhou, L.-D. Zhao, Adv. Mater. **29**(45), 1702676 (2017)
6. Y. Pei, X. Shi, A. LaLonde, H. Wang, L. Chen, G.J. Snyder, Nature **473**(7345), 66 (2011)
7. A. Banik, U.S. Shenoy, S. Anand, U.V. Waghmare, K. Biswas, Chem. Mater. **27**(2), 581 (2015)
8. S.V. Faleev, F. Léonard, Phys. Rev. B **77**(21), 214304 (2008)
9. L.D. Hicks, M.S. Dresselhaus, Phys. Rev. B **47**(19), 12727 (1993)
10. L.D. Hicks, T.C. Harman, X. Sun, M.S. Dresselhaus, Phys. Rev. B **53**(16), R10493 (1996)
11. M.S. Dresselhaus, G. Dresselhaus, X. Sun, Z. Zhang, S.B. Cronin, T. Koga, Phys. Solid State **41**(5), 679 (1999)
12. J.P. Heremans, V. Jovovic, E.S. Toberer, A. Saramat, K. Kurosaki, A. Charoenphakdee, S. Yamanaka, G.J. Snyder, Science **321**(5888), 554 (2008)
13. P.G. Klemens, Phys. Rev. **119**(2), 507 (1960)
14. M. Samanta, K. Biswas, J. Am. Chem. Soc. **139**(27), 9382 (2017)
15. K. Biswas, J. He, Q. Zhang, G. Wang, C. Uher, V.P. Dravid, M.G. Kanatzidis, Nat. Chem. **3**(2), 160 (2011)
16. K.F. Hsu, S. Loo, F. Guo, W. Chen, J.S. Dyck, C. Uher, T. Hogan, E.K. Polychroniadis, M.G. Kanatzidis, Science **303**(5659), 818 (2004)
17. K. Biswas, J. He, I.D. Blum, C.-I. Wu, T.P. Hogan, D.N. Seidman, V.P. Dravid, M.G. Kanatzidis, Nature **489**(7416), 414 (2012)
18. D.T. Morelli, V. Jovovic, J.P. Heremans, Phys. Rev. Lett. **101**(3), 035901 (2008)
19. S.N. Guin, A. Chatterjee, D.S. Negi, R. Datta, K. Biswas, Energy Environ. Sci. **6**(9), 2603 (2013)
20. G. Tan, F. Shi, S. Hao, L.-D. Zhao, H. Chi, X. Zhang, C. Uher, C. Wolverton, V.P. Dravid, M.G. Kanatzidis, Nat. Commun. **7**, 12167 (2016)
21. S.N. Guin, A. Banik, K. Biswas, Thermoelectric energy conversion in layered metal Chalcogenides, in *2d Inorganic Materials Beyond Graphene*, (World Scientific (WS), London, 2017), p. 239
22. Y. Sun, H. Cheng, S. Gao, Q. Liu, Z. Sun, C. Xiao, C. Wu, S. Wei, Y. Xie, J. Am. Chem. Soc. **134**(50), 20294 (2012)
23. T.C. Harman, P.J. Taylor, M.P. Walsh, B.E. LaForge, Science **297**(5590), 2229 (2002)
24. H. Böttner, G. Chen, R. Venkatasubramanian, MRS Bull. **31**(3), 211 (2006)
25. L.D. Hicks, M.S. Dresselhaus, Phys. Rev. B **47**(24), 16631 (1993)
26. D.A. Broido, T.L. Reinecke, Phys. Rev. B **51**(19), 13797 (1995)
27. B. Poudel, Q. Hao, Y. Ma, Y. Lan, A. Minnich, B. Yu, X. Yan, D. Wang, A. Muto, D. Vashaee, X. Chen, J. Liu, M.S. Dresselhaus, G. Chen, Z. Ren, Science **320**(5876), 634 (2008)

28. N.F. Mott, H. Jones, *The Theory of the Properties of Metals and Alloys* (Dover Publications, New York, 1958)
29. D.M. Rowe, *CRC Handbook of Thermoelectrics* (CRC press, Boca Raton, 1995)
30. J.P. Heremans, B. Wiendlocha, A.M. Chamoire, Energy Environ. Sci. **5**(2), 5510 (2012)
31. J.S. Son, M.K. Choi, M.-K. Han, K. Park, J.-Y. Kim, S.J. Lim, M. Oh, Y. Kuk, C. Park, S.-J. Kim, T. Hyeon, Nano Lett. **12**(2), 640 (2012)
32. G. Chen, J. Heat Transf. **119**(2), 220 (1997)
33. H.J. Goldsmid, R.W. Douglas, Br. J. Appl. Phys. **5**(12), 458 (1954)
34. M. Takashiri, S. Tanaka, K. Miyazaki, Thin Solid Films **519**(2), 619 (2010)
35. J. Chen, X. Zhou, C. Uher, X. Shi, J. Jun, H. Dong, Y. Li, Y. Zhou, Z. Wen, L. Chen, Acta Mater. **61**(5), 1508 (2013)
36. A.M. Adam, E. Lilov, P. Petkov, Superlattice. Microst. **101**, 609 (2017)
37. P. Nuthongkum, R. Sakdanuphab, M. Horprathum, A. Sakulkalavek, J. Electron. Mater. **46**(11), 6444 (2017)
38. C. Sudarshan, S. Jayakumar, K. Vaideki, C. Sudakar, Thin Solid Films **629**, 28–38 (2017)
39. M.-W. Jeong, S. Na, H. Shin, H.-B. Park, H.-J. Lee, Y.-C. Joo, Electron. Mater. Lett. **14**(4), 426 (2018)
40. Z.-K. Cai, P. Fan, Z.-H. Zheng, P.-J. Liu, T.-B. Chen, X.-M. Cai, J.-T. Luo, G.-X. Liang, D.-P. Zhang, Appl. Surf. Sci. **280**, 225 (2013)
41. P. Nuthongkum, A. Sakulkalavek, R. Sakdanuphab, J. Electron. Mater. **46**(5), 2900 (2017)
42. L.W. da Silva, M. Kaviany, Miniaturized thermoelectric cooler, in *ASME 2002 International Mechanical Engineering Congress and Exposition*, (American Society of Mechanical Engineers, New York, 2002), pp. 249–263
43. H. Zou, D.M. Rowe, S.G.K. Williams, Thin Solid Films **408**(1), 270 (2002)
44. L.W.d. Silva, M. Kaviany, C. Uher, J. Appl. Phys. **97**(11), 114903 (2005)
45. L.M. Goncalves, J.G. Rocha, C. Couto, P. Alpuim, M. Gao, D.M. Rowe, J.H. Correia, J. Micromech. Microeng. **17**(7), S168 (2007)
46. H. Bottner, J. Nurnus, A. Gavrikov, G. Kuhner, M. Jagle, C. Kunzel, D. Eberhard, G. Plescher, A. Schubert, K. Schlereth, J. Microelectromech. Syst. **13**(3), 414 (2004)
47. D.-H. Kim, E. Byon, G.-H. Lee, S. Cho, Thin Solid Films **510**(1), 148 (2006)
48. L.M. Goncalves, C. Couto, P. Alpuim, A.G. Rolo, F. Völklein, J.H. Correia, Thin Solid Films **518**(10), 2816 (2010)
49. G.J. Snyder, J.R. Lim, C.-K. Huang, J.-P. Fleurial, Nat. Mater. **2**, 528 (2003)
50. L.M. Gonçalves, The deposition of Bi_2Te_3 and Sb_2Te_3 thermoelectric thin-films by thermal co-evaporation and applications in energy harvesting, in *Thermoelectrics and its Energy Harvesting*, (CRC press, Boca Raton, 2012)
51. Z. Zhang, Y. Wang, Y. Deng, Y. Xu, Solid State Commun. **151**(21), 1520 (2011)
52. C.-L. Chen, Y.-Y. Chen, S.-J. Lin, J.C. Ho, P.-C. Lee, C.-D. Chen, S.R. Harutyunyan, J. Phys. Chem. C **114**(8), 3385 (2010)
53. H. Obara, S. Higomo, M. Ohta, A. Yamamoto, K. Ueno, T. Iida, Jpn. J. Appl. Phys. **48**(8R), 085506 (2009)
54. H. Choi, K. Jeong, J. Chae, H. Park, J. Baeck, T.H. Kim, J.Y. Song, J. Park, K.-H. Jeong, M.-H. Cho, Nano Energy **47**, 374 (2018)
55. M. Tan, Y. Hao, Y. Deng, D. Yan, Z. Wu, Sci. Rep. **8**(1), 6384 (2018)
56. M. Tan, Y. Hao, Y. Deng, J. Chen, Appl. Surf. Sci. **443**, 11 (2018)
57. F. Völklein, V. Baier, U. Dillner, E. Kessler, Thin Solid Films **187**(2), 253 (1990)
58. J. Tan, K. Kalantar-zadeh, W. Wlodarski, S. Bhargava, D. Akolekar, A. Holland, G. Rosengarten, Thermoelectric properties of bismuth telluride thin films deposited by radio frequency magnetron sputtering, in *Microtechnologies for the new millennium 2005*, (SPIE, Bellingham, 2005), p. 8
59. M. Tan, Y. Deng, Y. Wang, Nano Energy **3**, 144 (2014)
60. A. Foucaran, A. Sackda, A. Giani, F. Pascal-Delannoy, A. Boyer, Mater. Sci. Eng. B **52**(2), 154 (1998)

61. R.J. Mehta, Y. Zhang, C. Karthik, B. Singh, R.W. Siegel, T. Borca-Tasciuc, G. Ramanath, Nat. Mater. **11**(3), 233 (2012)
62. M.K. Jana, K. Biswas, C.N.R. Rao, Chem. Eur. J. **19**(28), 9110 (2013)
63. W. Wang, B. Poudel, J. Yang, D.Z. Wang, Z.F. Ren, J. Am. Chem. Soc. **127**(40), 13792 (2005)
64. G.-H. Dong, Y.-J. Zhu, L.-D. Chen, J. Mater. Chem. **20**(10), 1976 (2010)
65. G.-H. Dong, Y.-J. Zhu, L.-D. Chen, Cryst. Eng. Comm. **13**(22), 6811 (2011)
66. Y. Min, J. W. Roh, H. Yang, M. Park, S.I. Kim, S. Hwang, S.M. Lee, K.H. Lee, U. Jeong, Adv. Mater. **25**(10), 1425 (2013)
67. L.-D. Zhao, S.-H. Lo, Y. Zhang, H. Sun, G. Tan, C. Uher, C. Wolverton, V.P. Dravid, M.G. Kanatzidis, Nature **508**(7496), 373 (2014)
68. C. Chang, M. Wu, D. He, Y. Pei, C.-F. Wu, X. Wu, H. Yu, F. Zhu, K. Wang, Y. Chen, L. Huang, J.-F. Li, J. He, L.-D. Zhao, Science **360**(6390), 778 (2018)
69. L.-D. Zhao, C. Chang, G. Tan, M.G. Kanatzidis, Energy Environ. Sci. **9**(10), 3044 (2016)
70. G. Han, S.R. Popuri, H.F. Greer, J.-W.G. Bos, W. Zhou, A.R. Knox, A. Montecucco, J. Siviter, E.A. Man, M. Macauley, D.J. Paul, W.-g. Li, M.C. Paul, M. Gao, T. Sweet, R. Freer, F. Azough, H. Baig, N. Sellami, T.K. Mallick, D.H. Gregory, Angew. Chem. Int. Ed. **55**(22), 6433 (2016)
71. S. Chandra, A. Banik, K. Biswas, ACS Energy Lett. **3**(5), 1153 (2018)
72. S. Saha, A. Banik, K. Biswas, Chem. Eur. J. **22**(44), 15634 (2016)
73. Y. Luo, Y. Zheng, Z. Luo, S. Hao, C. Du, Q. Liang, Z. Li, K.A. Khor, K. Hippalgaonkar, J. Xu, Q. Yan, C. Wolverton, M.G. Kanatzidis, Adv. Energy Mater. **8**, 1702167 (2018)
74. Y. Ding, B. Xiao, G. Tang, J. Hong, J. Phys. Chem. C **121**(1), 225 (2016)
75. M.G. Kanatzidis, Acc. Chem. Res. **38**, 361 (2005)
76. M. Ruck, P.F. Poudeu Poudeu, Z. Anorg. Allg. Chem. **634**(3), 482 (2008)
77. M. Ohta, D.Y. Chung, M. Kunii, M.G. Kanatzidis, J. Mater. Chem. A **2**(47), 20048 (2014)
78. P.F. Poudeu, M.G. Kanatzidis, Chem. Commun. (21), 2672 (2005)
79. R. Atkins, M. Dolgos, A. Fiedler, C. Grosse, S.F. Fischer, S.P. Rudin, D.C. Johnson, Chem. Mater. **26**(9), 2862 (2014)
80. A. Mrotzek, M.G. Kanatzidis, Acc. Chem. Res. **36**(2), 111 (2003)
81. A. Mrotzek, M.G. Kanatzidis, J. Solid State Chem. **167**(2), 299 (2002)
82. L. Zhang, D. Singh, Phys. Rev. B **81**(24), 245119 (2010)
83. A. Chatterjee, K. Biswas, Angew. Chem. Int. Ed. **54**(19), 5623 (2015)
84. A. Banik, K. Biswas, Angew. Chem. Int. Ed. **56**(46), 14561 (2017)
85. L.E. Shelimova, O.G. Karpinskii, P.P. Konstantinov, E.S. Avilov, M.A. Kretova, V.S. Zemskov, Inorg. Mater. **40**(5), 451 (2004)
86. M.B. Babanly, E.V. Chulkov, Z.S. Aliev, A.V. Shevelkov, I.R. Amiraslanov, Russ. J. Inorg. Chem. **62**(13), 1703 (2017)
87. K. Nakayama, K. Eto, Y. Tanaka, T. Sato, S. Souma, T. Takahashi, K. Segawa, Y. Ando, Phys. Rev. Lett. **109**(23), 236804 (2012)
88. K. Yang, W. Setyawan, S. Wang, M.B. Nardelli, S. Curtarolo, Nat. Mater. **11**, 614 (2012)
89. M.G. Vergniory, T.V. Menshchikova, I.V. Silkin, Y.M. Koroteev, S.V. Eremeev, E.V. Chulkov, Phys. Rev. B **92**(4), 045134 (2015)
90. R. Vilaplana, J.A. Sans, F.J. Manjón, A. Andrada-Chacón, J. Sánchez-Benítez, C. Popescu, O. Gomis, A.L.J. Pereira, B. García-Domene, P. Rodríguez-Hernández, A. Muñoz, D. Daisenberger, O. Oeckler, J. Alloys Compd. **685**, 962 (2016)
91. A.M. Kusainova, P.S. Berdonosov, L.G. Akselrud, L.N. Kholodkovskaya, V.A. Dolgikh, B.A. Popovkin, J. Solid State Chem. **112**(1), 189 (1994)
92. S.D.N. Luu, P. Vaqueiro, J. Materiomics **2**(2), 131 (2016)
93. L.-D. Zhao, J. He, D. Berardan, Y. Lin, J.-F. Li, C.-W. Nan, N. Dragoe, Energy Environ. Sci. **7**(9), 2900 (2014)
94. M. Samanta, S.N. Guin, K. Biswas, Inorg. Chem. Front. **4**(1), 84 (2017)
95. L. Yang, Z.-G. Chen, G. Han, M. Hong, Y. Zou, J. Zou, Nano Energy **16**, 367 (2015)
96. H. Liu, X. Shi, F. Xu, L. Zhang, W. Zhang, L. Chen, Q. Li, C. Uher, T. Day, G.J. Snyder, Nat. Mater. **11**(5), 422 (2012)

Chapter 9
Thermoelectric Oxide Thin Films with Hopping Transport

Yohann Thimont

9.1 Introduction

In this chapter, we will focus our interest on semiconductors which are concerned by a hopping conduction for thermoelectric applications and more specifically with a thin film configuration. For the last years, the thermoelectric properties of oxides were studied to characterize these materials because the study of the Seebeck coefficient gives fundamental information about the Fermi level (E_F), effective mass (m^*), density of state (DOS), sign of possible structural transitions, and carrier density. In terms of thermoelectric performances, characterized by the power factor PF ($PF = \sigma\alpha^2$) where σ is the electrical conductivity and α the Seebeck coefficient and figure of merit ZT ($ZT = \sigma\alpha^2 T/k$ where k is the thermal conductivity of the material), their performances remain often lower in comparison with the commonly degenerated semiconductors intermetallic materials as silicides [1–3], tellure [4, 5], skutterudites [6, 7], Zintl [8], and half Heusler [9, 10]. The figure of merit of these materials is close to one and for some of them can be over 1 at their optimal temperature. Nevertheless, some oxides like $NaCo_2O_4$ show competitive figure of merit (also close to one) and the main oxides show other advantages in comparison with the previously cited materials such as their chemical stability in air, low toxicities, densities, and is relatively cheapness. It reveals also some interests for future specific TE applications and initiates new researches to increase the thermoelectric performances of oxide materials. A lot of oxide materials show semiconductor behaviors, their electrical conductivity increases with the temperature leading to potential technological applications as the transparent conducting oxides (TCO) such as In_2O_3, ZnO, and TiO_2 [11–13]. These oxides

Y. Thimont (✉)
CIRIMAT, Université de Toulouse, CNRS, Université Toulouse 3 Paul Sabatier,
Toulouse Cedex 9, France
e-mail: thimont@chimie.ups-tlse.fr

© Springer Nature Switzerland AG 2019
P. Mele et al. (eds.), *Thermoelectric Thin Films*,
https://doi.org/10.1007/978-3-030-20043-5_9

show a large band gap that can be used to avoid the photons absorption, thanks to an interband electronic transition corresponding to the visible range and needs also good electronics conductivities. Other common target applications are solar cells, electrochromic devices, chemical and temperature sensors, high temperature superconductors ($YBa_2Cu_3O_7$) [14], dielectrics ($SrTiO_3$ [15], $BaTiO_3$ [16, 17]), temperature sensors (Mn_2NiO_4) [18], and of course thermoelectrics like the well-known $NaCo_2O_4$ [19], $BiSrCoO$ [20] (whiskers), and $Ca_3Co_4O_9$ compounds. For a targeted application, the band gap, doping concentration, and stoichiometry must be optimized to obtain good thermoelectric properties. The thermoelectricity allows to convert the thermal energy into electrical energy and vice versa, in particular if the oxide shows a good Seebeck coefficient and have a low thermal conductivity. The semiconductor properties relate to the type of transport which can be different in various semiconductors. We can distinguish two types of conduction in all semiconductors. The first type is related to the band conduction mechanism with free carriers in an allowed band, the highly degeneracy semiconductor which has also free carrier conduction and electrical properties close to those of metals. Here, the free carriers are moving freely when a voltage is applied. It is concerning the case of various oxides as In_2O_3 and ZnO. The second case, which is the main purpose of this chapter, is related to the hopping conduction mechanism, which has no really free carriers, where the electrical conduction operates by a carrier hop from a localized state (where they are not free) to another one. It is the case of the mixed valence oxides such as $CuCrO_2$ and $NaCo_2O_4$. These two various types of conduction then modify also the temperature behavior of the Seebeck coefficient.

The material shaping is an important point for the thermoelectric performance optimization because it is necessary to have a high versatility for tuning the TE material to specific applications in the aim to reduce the heat losses. TE materials deposited as thin films offer also the possibility to enhance ZT values due to specific microstructures which can lead to a significant reduction of the thermal conductivity compared to that observed in bulk materials [21–24]. Thin film geometry allows to have a better integration in heat dissipation devices. Nevertheless, microstructures obtained in the case of thin films can reduce also the electrical conduction of the film by free carrier scattering effects and the influence of the interfaces made also free carrier scattering. In the case of hopping semiconductors, the hop distances are clearly smaller than the film thickness, which leads to an advantage in terms of transport for the very thin thickness.

In the first part, we will discuss the properties of the thermoelectric properties based on hopping semiconductors and the case of thin films for thermoelectric applications. A second part will be devoted to the measurements of the transport properties of thin films. The last part of this chapter will be devoted to the thin films made with hopping semiconductors. We will show the case of the Mg doped $CuCrO_2$ compound oxide concerned by the hopping transport.

9.2 Thermoelectric Properties of the Thermoelectric Hopping Oxides

Hopping operates between two different oxidations states of the same elements (M) localized in the same type of site in the crystallographic structure. It is then possible to write the redox equations (where A and B are two similar and neighboring sites):

$$M_{(A)}^{(n+1)+} + e^- \iff M_{(A)}^{n+} \quad \text{and} \quad M_{(A)}^{n+} + h^+ \iff M_{(A)}^{(n+1)+}$$

The global equation is:

$$M_{(A)}^{n+} + M_{(B)}^{(n+1)+} \iff M_{(A)}^{(n+1)+} + M_{(B)}^{n+}$$

The hopping conduction is characterized by a carrier hop from a localized state to another. In this case, the electrical conductivity increases with the temperature not due to an increasing of the carrier concentration but due to the hop success which increases with the temperature. To have this type of conduction it is necessary to have element with a mixed valence in the material in a same type of structural site. This type of conduction was identified by Mott [25] in some materials.

This transport mode is often found in the case of the disordered correlated systems and the mixed valence materials. It is the case of a large number of oxide materials which have a spinel ($Mn_{3-x}Ni_xO_4$) [18, 26] and delafossite ($CuCrO_2$) structures [27], for instance, and also in some conducting polymers like p-PEDOT [28].

This mode of transport plays an important role in the thermoelectric properties. In fact, the Seebeck coefficient relates to the carrier concentration in the semiconductor. In the case of the hopping semiconductor for which the carrier concentration stays constant with the temperature, it is leading to a constant Seebeck coefficient with the temperature while the electrical conductivity increases. The relation between the Seebeck coefficient (α) and the mixed valence of the ions M concerned by the hopping is given in the simplest case by the Heikes formula (9.1):

$$\alpha = \frac{k_B}{q} \ln \left(\frac{M^{\alpha+}}{M^{\beta+}} \right) \tag{9.1}$$

As an example, Fig. 9.1 shows the Seebeck coefficient and the electrical conductivity variations with the temperature of the hopping semiconductor and is compared to other types of conductors.

The sign of the Seebeck coefficient gives an information about the semiconductor transport type. In fact, p- and n-type hopping semiconductors are identified by the entropy flow (oriented from the most ordered state to the most disordered state for a positive entropy flow) in the structure (see Fig. 9.2 in the example of the M^{n+} normal state cation) and not by the nature of the carrier (electron and hole). In both cases, when electron hops, a hole hops too but in the opposite direction.

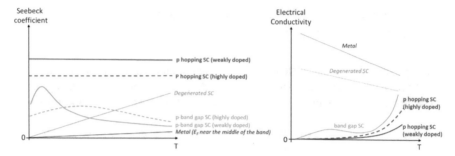

Fig. 9.1 *Left*: Schematic variation of the Seebeck coefficient with the temperature for various types of materials (hopping semiconductor with a constant carrier concentration and no variation of the spin-orbital degeneracy, band gap semiconductors with free carriers, degenerated semiconductors and metals). The variation of the Seebeck coefficient in the band gap semiconductor with two doping concentrations is inspired by Snyder et al. [29]. *Right*: Schematic variation of the electrical conductivity of these various types of materials

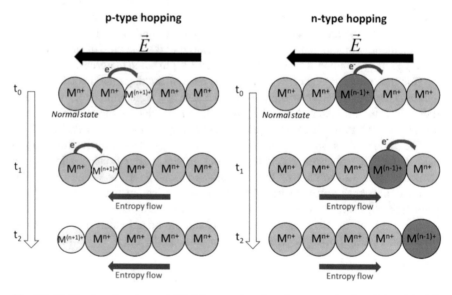

Fig. 9.2 (*Left*) p-type hopping and (*Right*) n-type hopping along a conducting chain with their corresponding entropy flow in an M^{n+} matrix (the ions radii are correlated to their respective oxidation degrees and the positive entropy flow schematized by arrows)

There are many other oxide materials which are concerned by the extended Heikes relation like $CaMnO_3$, $Ca_3Co_4O_9$, $CaMn_{3-x}Cu_xMn_4O_{12}$ or $LiMn_2O_4$, and $Fe_xMn_{1-x}NiCrO_4$ spinel compounds.

The g_1/g_2 ratio drives to the entropy flow which is at the origin of the p- or n-type transport nature whatever the carrier type. The net entropy flow is given by (9.2):

$$S_{\text{net}} = k_B \ln \frac{g_1}{g_2} \qquad (9.2)$$

It is possible to enhance the Seebeck coefficient, thanks to the g_1/g_2 ratio in some structures. As an example, Kobayashi et al. [30] have showed the influence of the g_1/g_2 ratio on the Seebeck coefficient in the case of the $CaMn_{3-x}Cu_xMn_4O_{12}$ compound.

A study was led on $NaCo_2O_4$ oxides by Terasaki et al. [19]. This oxide shows a metallic electrical resistivity of (0.2 mΩ cm) [31] and its Seebeck coefficient increases from $+100$ μV/K at 300 K to 200 μV/K at 800 K which drives to a ZT near 1 at 800 K. To this day, it is the oxide compound which shows the highest thermoelectric performances. This compound shows a high Seebeck coefficient with a large carrier density (10^{22} cm^{-3}) and the Seebeck coefficient increases with the temperature. Its high magnitude is due to a high g_1/g_2 ratio which is equal to 1/12 [32]. This large ratio is due to the possibility to have simultaneously high spin and low spin states in the compound leading to a high global entropy per carrier. The high and low spin proportions vary with the temperature, which explains the variation of the Seebeck coefficient and the electrical conductivity with the temperature in the case of such compound.

Moreover, in the case of hopping transport, their thermal conductivities remain relatively low (it is depending on their structures) because of the conduction heat transfer which operates mainly by the phonons and not by the carriers which are not free. The Wiedemann–Franz relation [33] application is not well appropriate in this case. It was demonstrated in the case of the VO_2 compound, for example. Although their electrical conductivities are relatively low due to the transport behavior, the oxides concerned by hopping transport are studied for various applications and in particular for thermoelectric applications. Moreover, a lot of these oxide materials are ecofriendly, cheap, and light. Some researches are also devoted to use this kind of semiconductors in thin film configurations for specific thermoelectric applications.

9.3 Thin Films for Thermoelectric Applications

9.3.1 Thin Films Preparation and Specificity

To this day, there are only a few thermoelectric applications which are based on thin films of some hundred nanometers in comparison with bulk materials. It is due to their low electrical power generation. Nevertheless, thin films have other advantages like, as shown below.

Thin films are in adequacy with the miniaturization of the thermoelectric devices and are also suitable for fitting to the geometry of the waste heat dissipation devices (which reduces the thermal loss) or as accurate temperature probes (due to a low thermal pumping and good thermal contact).

Thin films can be elaborated by physical vapor deposition techniques which can be sputtering and pulsed laser deposition (PLD) techniques or by chemical vapor deposition techniques as atomic layer deposition (ALD). In the case of sputtering techniques, a plasma of inert gas is generated and ions will bombard a target. A kinetic energy of the ions in the plasma will be exchanged with the atoms of the target. These atoms will be extracted from the target and will be deposited on a substrate. In the case of PLD, a laser is used to extract atoms from the target. In both cases, the chemical compositions of the films are similar to those of the targets. Otherwise, in the case of the CVD process, a chemical reaction is initiated at the surface of the substrate with precursors. Due to the deposition techniques, the microstructure of the films can be very different in comparison with the bulk.

The crystallization of the film can be obtained after the deposition step or after post-annealing treatment. Films can be single crystals or in polycrystalline. The influence of the crystallinity of the substrate and its crystal orientation is very important for the film growth and for the transport properties. Seebeck coefficient relates to the band structure and can be anisotropic. Thin films can be orientated according to a crystallographic orientation (preferred orientation) and it is depending on the substrate. The orientation can be profitable for transport properties.

In an additional point, thin films made by PVD route can show specific nanostructuration as a function of the deposition parameters [34]. The nanostructuration allows to have higher phonon scattering effects which can strongly reduce the thermal conductivity of the film material. The influence of the nanostructuration or grains with nanometric sizes was strongly studied by Dresselhaus et al. [35] which have led to a large benefit on the ZT.

The nanostructuration and stress effect induced by the substrate can also impact the band structure of the deposited crystal material and consequently the density of state which has a large influence on the Seebeck coefficient of band gap and degenerated semiconductors with free carriers.

In the case of thin films, the grains can be smaller than in a bulk material. When the grains become very small, the quantum confinement can be present, which increases consequently the band gap energy and is then profitable for high temperature thermoelectric applications because it reduces the presence of both carrier types.

The deposition techniques make possible the stabilization of metastable compounds such as the spinel oxide $Co_{1.7}Fe_{1.3}O_4$, while the thermodynamic laws lead to a spinodal decomposition [36] in two different phases ($Co_{1.16}Fe_{1.84}O_4$ and $Co_{2.7}Fe_{0.3}O_4$) according to the phase diagram which could initiate new interest for this compounds.

The influence of the substrate on the thermal properties gives the possibility to tune the thermal conductivity of the system.

The last point concerns the heat flux exchange due to a high surface/thickness ratio. The emissivity and convection have a stronger impact in the case of thin films and they can be suitable for thermoelectric applications due to the possibility to have higher temperature gradients in the plane of the film. Nevertheless, the emissivity of the film is rarely taken into consideration in the case of thermoelectric applications and it should be more studied.

Due to these various effects, thin films show a large versatility for thermoelectric applications. We will explain some properties of thermoelectric material when they are deposited as thin films.

9.3.2   Thin Films Geometry for Thermoelectric Applications

Two types of thin films geometries can be used for thermoelectric applications. Geometries can be identified according to the heat flux direction q_h as it is schematized in Fig. 9.3.

The planar geometry allows to obtain high temperature gradients (only if the substrate and films have not too high thermal conductivities) but shows small current section (film width multiplied by film thickness). This type of geometry allows to have a high voltage in an open circuit (E_{oc}) but with low power production.

The electrical power (P) produced by the film can be approximated by 9.3 (the contact resistances are neglected):

$$P = E_{oc}I - rI^2 \tag{9.3}$$

where:

$$E_{oc} = -\alpha \times \Delta T \tag{9.4}$$

and

$$r = \frac{\rho L}{S} \tag{9.5}$$

(with S the section through by the current and ρ the resistivity of the material).
Then:

$$P = -\alpha \Delta T \times I - \frac{\rho L}{S} \times I^2 \tag{9.6}$$

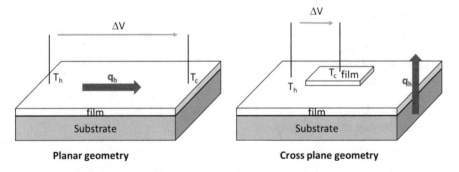

Fig. 9.3 The two geometries used for TE applications in the case of thin films: planar and cross plane geometry

By definition:

$$R_{\text{sheet}} = \frac{\rho}{t} \qquad (9.7)$$

with t the thickness of the film.

Then the electrical power (P) is equal to (9.8):

$$P = -\alpha \Delta T \times I - R_{\text{sheet}} \times F \times I^2 \qquad (9.8)$$

with F the shape factor which depends on the film size.

- In the case of a planar geometry, the first term ($-\alpha \Delta T \times I$) can be high because of high temperature gradient which can be easily obtained but the second term is also high because of the low thickness which can increase drastically the electrical resistivity when the thickness is in the range order of the carrier mean free path (λ). The main problem of this kind of geometry is the high electrical resistance which decreases the electrical power. This geometry was used in various works for thermoelectric generators as described by Fang et al. [37] and Zappa et al. [38].

  With this planar geometry, there is a large influence of the substrate on the global temperature gradient because of the large contribution of substrate thermal conductivity. It is similar to parallel thermal resistances. In this case, it is possible to use, in some conditions (thickness of the film, emissivity of the film, and thermal conductivity of the film), a modified figure of merit $(ZT)^*$ of the composite (film and substrate) where the thermal conductivity of the thermoelectric material is substituted by those of the substrate ($k_{\text{substrate}}$) [39] (9.9)

$$(ZT)^* = \frac{\sigma \alpha^2}{\kappa_{\text{substrate}}} T \qquad (9.9)$$

  In this case, the (ZT*) of the system film and substrate becomes higher than the ZT of thermoelectric in particular if the substrate have a lower thermal conductivity than those of the thermoelectric material. Sinnarasa et al. [40] have modeled the influence of the thickness on the temperature difference in the case of CuCrO2 thin films (5 W/m/K) deposited on fused silica substrates (ksubstrate = 1.38 W/m/K). They have shown that the influence of the film on the thermal properties can be neglected for a thickness below 1 μm in this case. Finite element models are necessary to determine the limit of validity of the (ZT)* and need to take into consideration the film emissivity.

- The cross plane geometry (it is analogous to the conventional TE modules geometry) in opposite has a small temperature gradient (along a step of some 100 nanometers). Then the first and the second terms become very low. The main problem of this geometry is due to the temperature gradient which remains small and leads to a low E_{oc}. Then the electrical power remains low.

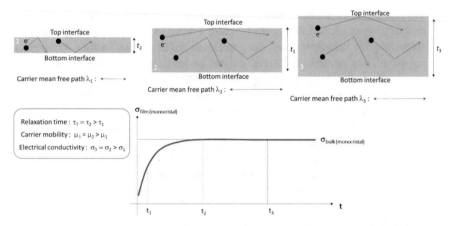

Fig. 9.4 Representation of the mean free path of free carriers in a thin film and the variation of the electrical conductivity with the film thickness

As a conclusion, to obtain high powers, it is necessary to have the lower resistance as possible (need a good optimization of the thin film thickness and nanostructuration) and the possibility to have the highest temperature gradient as possible. It is difficult to have both, then it is the reason why the thermoelectric thin films (module) are not well adapted for high electrical power production but can be used for microelectronic power production (μWatt) or also as temperature sensors.

The film thickness plays a large role in the electrical conductivity in the case of compounds with free carrier transport (in section 9.3.2). In the case of thin films, for which the thickness (t) is in the range order of the carrier mean free path (λ), a film thickness decreasing below λ leads to a lower electrical conductivity (σ) because of the carrier scattering at the interface and the surface which reduces the relaxation time (see Fig. 9.4). The film thickness must be optimized as a function of the carrier mean free path in this case.

The nanostructuration and the microstructure of films impact the electrical conductivity when the grain sizes are in the range order of the carrier mean free path. Then it is necessary to have a value of the carrier mean free path length different from those of the phonons wavelength in order to diffuse only phonons without any impact on the carrier transport.

9.3.3   The Case of the Hopping Transport in Thin Films for Thermoelectric Applications

In the case of the hopping transport in the thermally activated mode, the hop lengths are in the range order of the lattice parameters. These distances are very small in comparison with the thickness of the film or grain sizes. In this case, we cannot consider a carrier scattering effect by interface.

- We can conclude that the film thickness does not impact the intrinsic hopping mobility because the film thickness is larger than the average hopping distance.
- Regarding the microstructure, the grain sizes are almost larger than the average hopping distance and the carrier cannot be scattered by it. Nevertheless, the grain boundaries can play an important role due to its relatively low width that makes the carrier going through (it is also relative to the electronic wave function amplitude damping of the carrier for the tunnel effect). The carrier percolation is also required between grains to have a global current transport.
- In opposition, the stress induced by the substrate can have an impact on the hopping frequency and the hopping success probability because stress makes an interatomic distance variation (lattice deformation). For a low damping of the wave function amplitude, if the stress induces an increase of the interatomic distance along the hop direction, then the carrier mobility increases too. If the damping dominates, then the mobility decreases.
- For oriented crystallized thin films, it is possible to have a large impact on the electrical conductivity as a function of the film orientation. As an example, if the hopping operates in the (ab) plane of the crystallographic structure, $\vec{c}$ axis growth will be a benefit for the global electrical conductivity.

The mobility can be similar to those of the bulk material with hopping transport in the case of thin films with small grain boundary width and if no stress is present in the film (which can impact the average hop distance). These characteristics generate some interests for thin films with hopping transport because the difference of mobility between the bulk and the thin film can be relatively low in comparison with the free carrier conduction mode encountered in metal and band gap semiconductors. The mobility remains often relatively small and leads to a low electrical conductivity in the case of a lot of oxides materials. However, some hopping semiconductor compounds like $NaCo_2O_4$, $Ca_3Co_4O_9$, and $Bi_2Sr_2Co_2O_9$ misfits show very good electrical conductivities and Seebeck coefficients and should be more studied because they have also low thermal conductivities due to their misfit structure (with misalignment of the octahedron layers).

Due to their possible constant Seebeck coefficient (when there are no variation of the carrier concentration due to the stoichiometry or spin-orbital degeneracy), the hopping oxide thin films can mainly be used as accurate temperature probes because it is not necessary to take into consideration the variation of the Seebeck coefficient with the temperature by any electronic devices leading to a cheaper and smaller probes.

9.4 Measurement of the Physical Properties of Hopping Conduction Thin Films (in Plane Geometry)

Measurements of the physical properties of thin films require good electrical contacts. Therefore, metal electrodes can be deposited on top of the film or metallic wires can also be bonded directly on the surface of the film using a wire bonder apparatus. The electrical conductivity measurement is obtained by the four probes technique. Four probes are in the equivalent contact with the top of the film. The current is injected into the film by a first probe and collected by the fourth probes then the voltage is measured, thanks to the two medium probes. A voltage is read when the current is applied then a resistance can be determined. It is necessary to check the linearity of the voltage versus the current curve with various current values injection. This condition is necessary to check the ohmic contact nature only (avoid the Schottky contact). If the contact is a Schottky type, it is necessary to change the metallic nature of the probes (have a low extraction potential metal for the characterization of an n-type material as an example). The resistivity can be calculated from the value of the resistance by applying shape correction factors and using the film thickness. The calculations were detailed by Smith et al. [41]. The determination of an accurate value of the Seebeck coefficient in thin films was described by Bahk et al. [42].

To apply a thermal gradient, it is necessary to have a good thermal contact between the thin film/substrate and the heater (silver paste can be used). It is necessary to have only one temperature gradient in the film/substrate. It is the main problem of the two heater plates technique. In this case, when the film/substrate has a low thermal conductivity, an intermediate location (located between the two heaters) can show a lower temperature than the cold side leading to an error on the Seebeck coefficient determination if the Seebeck coefficient depends on the temperature. The heater in a furnace technique is commonly used for bulk materials but it is applicable in the case of thin films only if the thermal contacts are good. Figure 9.5 shows the temperature gradient in the plane of the film/substrate for both measurement techniques and in the case of a high and low film/substrate thermal conductivity.

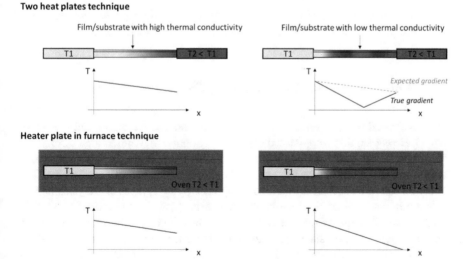

Fig. 9.5 Representation of the temperature gradient in the film/substrate for both Seebeck measurement techniques on the films/substrate with high (*Left*) and low (*Right*) thermal conductivity

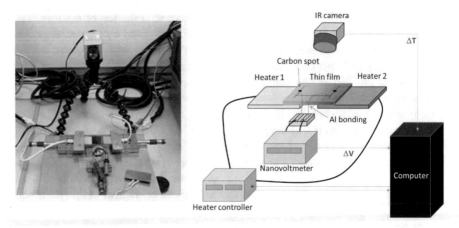

Fig. 9.6 (*Left*) Photography of a homemade setup for the measurement of an oxide thin film deposited on a fused silica substrate. (*Right*) Schematic representation of the setup

In addition, the electrical contact and the temperature measurement must be performed on the same isotherm line and are often at the origin of some errors.

The determination of the temperature at the surface is difficult, because the thermocouple pumps heat from the films and introduces a bias. Infrared camera is a good alternative path because there is no heat pumping. Nevertheless, it is necessary to know the film emissivity which can be measured, thanks to an emissometer. It is also possible to use a carbon spot coated on top of the film to have a good temperature reading because carbon materials have a very high thermal conductivity and an emissivity close to one (black body) which does not vary much with the temperature.

As an example, a specific homemade Seebeck coefficient measurement setup (two heat plates) adapted for thin film geometry is represented in Fig. 9.6.

9.5 The Example of the Delafossite Oxide Thin Films

Charles Friedel was a chemist and mineralogist who discovered the $CuFeO_2$ compound. He named this compound delafossite in honor to the mineralogist Delafosse [43]. The ABO_2 oxide materials have a delafossite structure. In this structure, A is in dumbbell coordinance (2) and the B site is in octahedral coordinance. The structure can be described as A lattice plan joined by two planes of BO_6 octahedral. The oxidation state of the ions in the A site is $+I$ while the oxidation state is $+III$ for the B ions. When the B ions are substituted by other elements which have a lower oxidation state ($+II$), the mixed valence appears in the A site where both oxidations states ($+I$ and $+II$) can be found. In the case of the Mg doped $CuCrO_2$ delafossite, Mg^{2+} can substitute the Cr^{3+} and generate a mixed valence of the copper (Cu^+/Cu^{2+}) in the A site. The electrical conductivities found in this material family (doped bulk $CuMO_2$) vary from 1×10^{-4} S/cm to 30 S/cm at room temperature [44]. This material family shows good properties as p-type transparent conducting oxide thin films, also for photocatalysis, watersplitting, HCl oxidation [45], and thermoelectrics at high temperatures due to their large optical band gaps. The thermoelectric properties of undoped delafossite bulk materials were performed by Ruttanapun et al. [46], Mg doped bulk materials by Maignan et al. [47], and Mg doped thin films by Sinnarasa et al. [26]. Other models were also developed to explain the value of the Seebeck coefficient as the Kubo formalism [45].

9.5.1 Influence of the Annealing Temperature

Mg doped $CuCrO_2$ thin films were deposited by RF magnetron sputtering on fused silica substrate annealed under vacuum by Sinnarasa et al. [26]. They studied the influence of the annealing temperature (AT) and film thickness on the transport properties (electrical conductivity and Seebeck coefficient). Figure 9.7 reports the electrical conductivity and Seebeck coefficient of $CuCrO_2$ thin films as a function of temperature for various annealing temperatures.

Firstly, all Mg doped $CuCrO_2$ thin films show a positive Seebeck coefficient which confirms the p-type semiconducting transport. Moreover, the Seebeck coefficient does not vary with the measuring temperature while the electrical conductivity increases, which confirms the hopping transport mode with a hole concentration which does not depend upon the temperature in the delafossite material.

On the other hand, the electrical conductivity increases with the annealing temperature until 500 °C and remains constant. The opposite effect is showed in the case of the Seebeck coefficient and can be explained by the fact that the electrical conductivity and Seebeck coefficient are oppositely correlated by the

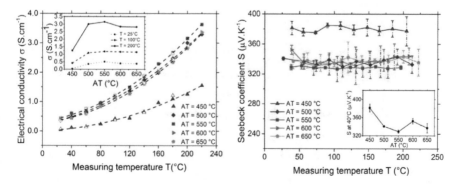

Fig. 9.7 Electrical conductivity (*Left*) and Seebeck coefficient (*Right*) as a function of the measuring temperature for various annealing temperatures (AT). Reproduced from open source reference [27]

carrier density. It is possible to conclude that the carrier density varies with the annealing temperature and remains constant for annealing temperature which is above 500 °C.

The mixed valence ratio $[Cu^+]/[Cu^{2+}]$ has been calculated from the extended Heikes formula which takes into consideration the copper environment and the spin-orbital degeneracy. In this case, the extended Heikes formula developed by Koshibae et al. [48] becomes (9.10):

$$\alpha = \frac{k_B}{q} \ln\left(\frac{g_1}{g_2} \frac{[Cu^+]}{[Cu^{2+}]}\right) \tag{9.10}$$

where g_1 and g_2 are the electron degeneracy factors for an oxidation state (1) or (2) regarding both atoms which participate in the hopping conduction. The electron degeneracy is equal to $g = (2S + 1)N$ where S is the global spin and N the number of system configurations.

Figure 9.8 shows the case of the delafossite $CuCrO_2$ with the orbital degeneracy of the copper in the dumbbell environment due to the O-Cu-O bond. In this case, the five 3D orbitals of the copper are split on three various energy levels (2 eg, 1 a1g, and 2 e'g). Due to the Mg^{2+} substitution (from Cr^{3+}), the copper can be found as Cu^+ (normal state) and Cu^{2+} in the copper plane which corresponds to a p-type hopping conduction, the energy level filled by electrons (3d) is then described in Fig. 9.7. The Cu^+ case shows only one possible configuration because all energy levels are

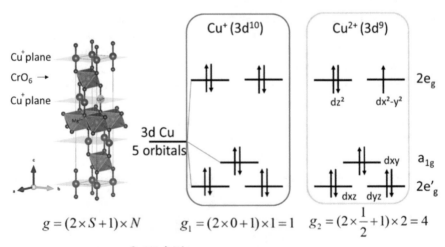

$$g = (2 \times S + 1) \times N \qquad g_1 = (2 \times 0 + 1) \times 1 = 1 \qquad g_2 = (2 \times \frac{1}{2} + 1) \times 2 = 4$$

S : total spin
N: number of possible configurations for a given energy

Fig. 9.8 Spin and orbital degeneracies of Cu^+ and Cu^{2+} in the delafossite structure Mg doped $CuCrO_2$

filled with two electrons with opposite spins according to the Pauli's exclusion. The Cu^{2+} ion shows two possible electronic configurations.

In the case of the film which shows the highest electrical conductivity (which is the sample annealed at 550 °C), the calculated Cu^{2+} fraction is equal to 0.005. Thanks to the lattice volume determined by XRD, the calculated hole density is then equal to 1.17×10^{20} cm^{-3}. This value also corresponds to the carrier density of thin films annealed at higher temperature than 550 °C because the Seebeck coefficient remains almost constant for thin films annealed at higher temperatures. Maignan et al. [49] have reported a magnesium substitution limit of some percent which is also in agreement with the exposed results. The 500 °C annealed thin film shows a lower hole density of only 1.32×10^{19} cm^{-3} which can be explained by an uncompleted substitution of the chromium by the magnesium due to an insufficient temperature during the annealing and an uncompleted crystallization of the $CuCrO_2$ phase. Combining the electrical conductivity and the hole density (determined by the Seebeck coefficient), the hole mobility has been calculated according to (9.11):

$$\mu = \frac{\sigma}{n\,|q|} \tag{9.11}$$

In the case of the $CuCrO_2$, the mobility remains less than 0.1 cm^2 V^{-1} s^{-1} which are in agreement with the Van Daal's limits [50] and justify an effective hopping transport in this compound.

9.5.2 Influence of the Film Thickness

The influence of the film thickness on the Seebeck coefficient can be also discussed in terms of thermoelectric properties. In the case of homogeneous film with hopping transport, the Seebeck coefficient should not depend on the film thickness. In fact, the hopping distance is very low in comparison with the film thickness and there is no effect of interface scattering which could influence the transport properties. Nevertheless, the influence of strain induced by the substrate and microstructure (by grain boundaries) can limit the doping substitution resulting to a variation of the carrier concentration and carrier mobility. These effects were identified in the case of the optimized annealed (at 550 °C) Mg doped $CuCrO_2$ thin films.

Fig. 9.9 shows the variation of the electrical conductivity and Seebeck coefficient with the film thickness at room temperature.

Firstly, the electrical conductivity increases until a thickness value of 100 nm then it remains constant between 100 and 400 nm and decreases drastically for a higher thickness (>400 nm). The interface scattering cannot explain the electrical behavior for the lower thickness because the transport is hopping type. The Seebeck coefficient gives more information which can explain the electrical properties. The Seebeck coefficient remains independent of the measuring temperature (not showed here). The lower thickness thin films show the highest Seebeck coefficient

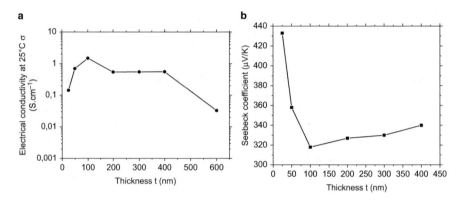

Fig. 9.9 (*Left*) Electrical conductivity as a function of the thickness of the Mg doped CuCrO$_2$ thin films. (*Right*) The Seebeck coefficient (at room temperature) as function of the thicknesses

which corresponds to a lower hole density. The glow discharge optical emission spectroscopy (GDOES) technique was employed to determine the magnesium distribution in the thickness of the films and revealed that the magnesium is segregated close to the interface in the case of the thinner films and becomes homogeneous for higher thicknesses (higher than 50 nm). The non-effective substitution of the chromium by the magnesium leads to a global carrier concentration decreasing in the case of the thinner films. The segregation of the magnesium could be caused by the stress induced by the substrate on the delafossite structure. XRD analysis revealed lattice parameter variation with the film thickness.

The high variation of the electrical conductivity for higher thickness (>400 nm) is not explained by the variation of the carrier density because the Seebeck coefficient remains constant, but a microscope observation revealed microcracks which avoid the current percolation and decreases drastically the electrical conductivity. The presence of microcracks reveals an important effect of strain induced in the film by the substrate.

9.5.3   Elaboration and Properties of Thin Film Thermoelectric Modules Containing a Hopping Oxide

Thin film thermoelectric modules with hopping transport oxides can be elaborated, thanks to a deposition process like the sputtering technique. Lift off, lift on, and shadow mask techniques can be employed to elaborate the TE strips (corresponding to legs). We can show the case of uni-strip (unileg) p-type thin film thermoelectric module with a planar geometry made with optimal annealed (550 °C) 100 nm Mg doped CuCrO$_2$ strips. Figure 9.10 (Left) shows a picture of the thermoelectric module with three Mg doped CuCrO$_2$ strips. Figure 9.10 (Right) shows the microstructure of the Mg doped CuCrO$_2$ strip. The film is dense as we can see on this micrography. Each 100 nm thick CuCrO$_2$ strip is electrically connected

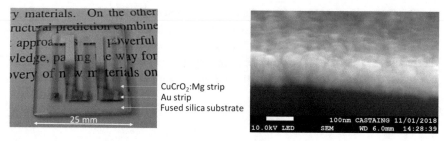

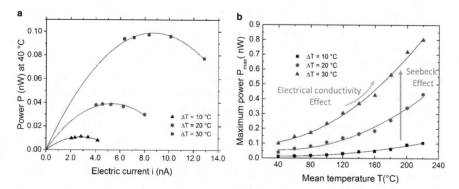

Fig. 9.10 (*Left*) Uni-strip thermoelectric thin film. (*Right*) Tilted SEM cross section of 100 nm Mg doped CuCrO$_2$ thick film

Fig. 9.11 (*Left*) Electrical power of the uni-strip Mg doped CuCrO$_2$ thermoelectric thin films as a function of the current for various temperature differences. (*Right*) Maximum of the electrical power as a function of the average temperature for various temperature differences

to gold strips (deposited by shadow mask technique after the annealing of the CuCrO$_2$ compound) and the electrical contact between gold and the p-type CuCrO$_2$ compound is of ohmic type. We can also notice that CuCrO$_2$ strips are transparent in the visible range which can be profitable for specific applications.

Applying a controlled thermal gradient along the strip, electrical powers were measured for various temperature differences and loading resistances. Figure 9.11 reports the electrical power generated as a function of the electrical current through the TE module and the maximum of the electrical power as a function of the mean temperature for various temperature differences.

In the case of a free thermal gradient, the temperature difference becomes more important because of the low thermal conductivity of the substrate (1.38 Wm^{-1} K^{-1} for the fused silica) and can reach easily more than 200 °C leading to a higher electrical power generation. As an example, for a 225 °C temperature difference, the 100 nm Mg doped CuCrO$_2$ uni-strip thermoelectric module with only three strips reaches a power of 11 nW. The performance of the device is pretty good in comparison with other thin film thermoelectric modules made with other oxide semiconductors as Al doped ZnO and Ca$_3$Co$_4$O$_9$. As an example, Saini et al. [51] obtained 30 pW for n-type Al doped ZnO (band semiconductor) and p-type

$Ca_3Co_4O_9$ (hopping semiconductor) strips in plane thin film thermoelectric modules for which the oxide was deposited on Al_2O_3 substrate and with an applied temperature difference of 230 °C. Although the power generated by the thin film thermoelectric modules is low in comparison with their bulk analog modules, they can generate some interest for microelectronic energy conversion which does not need high power supply.

9.6 Conclusion

In this chapter, we have discussed about the transport properties of the hopping semiconductors and cited the case of the various compounds such as $Ca_3Co_4O_9$, $NaCo_2O_4$, and $CuCrO_2$. These compounds can be deposited as thin films by physical vapor deposition techniques to make some specific thermoelectric devices. We have showed that specific oxide thin films with hopping transport and high band gap energy can be interest in terms of thermoelectrics and more particularly for high accurate temperature measurement sensors due to their very high stability (in temperature and atmosphere) and their Seebeck coefficient which not depend of the temperature. In terms of power supply, the thin film thermoelectric modules made with this kind of compounds show a performance which could be interesting for microelectronic energy conversion. Some of these compounds show additional properties as optical transparent materials that can add supplementary functions as smart windows which incorporate temperature sensors, for example.

Acknowledgement The author would like to thank Dr. C. Tenailleau for his contribution in the editing of the document.

References

1. W. Wunderlich, Y. Suzuki, N. Gibo, T. Ohkuma, M. Al-Abandi, M. Sato, A.U. Khan, T. Mori, Thermoelectric properties of Mg_2Si produced by new chemical route and SPS. Inorganics **2**(2), 351–362 (2014)
2. D. Berthebaud, F. Gascoin, Microwaved assisted fast synthesis of n and p-doped Mg_2Si. J. Solid State Chem. **202**, 61 (2013)
3. M.I. Fedorov, G.N. Isachenko, Silicides: materials for thermoelectric energy conversion. Jpn. J. Appl. Phys. **54**, 7S2 (2015)
4. J.P. Fleurial, A. Borshchevsky, M.A. Ryan, W.M. Phillips, J.G. Snyder, T. Caillat, E.A. Kolawa, J.A. Herman, P. Mueller, M. Nicolet, Development of thick-film thermoelectric microcoolers using electrochemical deposition, https://ntrs.nasa.gov/search.jsp?R=200000488092018-09-12T12:50:14+00:00Z
5. J.H. Kim, J.Y. Choi, J.M. Bae, M.Y. Kim, T.S. Oh, Thermoelectric characteristics of n-type Bi2Te3 and p-type Sb2Te3, thin films prepared by co-evaporation and annealing for thermopile sensor applications. Mater. Trans. **54**(4), 618–625 (2013)

6. G. Rogl, P. Rogl, Skutterudites, a most promising group of thermoelectric materials. Curr. Opin. Green Sustain. Chem **4**, 50–57 (2017). https://doi.org/10.1016/j.cogsc.2017.02.006
7. J.Q. Guo, H.Y. Geng, T. Ochi, S. Suzuki, M. Kikuchi, Y. Yamaguchi, S. Ito, Development of skutterudite thermoelectric materials and modules. J. Electron. Mater. **41**(6), 1036–1042 (2012)
8. S.M. Kauzlarich, S.R. Browna, G. Jeffrey Snyder, Zintl phases for thermoelectric devices. Dalton Trans. **21**, 2099 (2007)
9. S. Chen, Z. Ren, Recent progress of half-Heusler for moderate temperature thermoelectric applications. Mater. Today **16**(10), 387–395 (2013)
10. J. Yu, K. Xia, X. Zhao, T. Zhu, High performance p-type half-Heusler thermoelectric materials. J. Phys. D. Appl. Phys. **51**, 113001 (2018)
11. R. Ismail, O. Abdulrazzaq, Z.K. Yahya, Preparation and characterization of In2O3 thin films for optoelectronic applications. Surf. Rev. Lett. **12**(04), 515–518 (2005). https://doi.org/10.1142/s0218625x05007359
12. S. Agrawal, R. Rane, S. Mukherjee, in *ZnO Thin Film Deposition for TCO Application in Solar Cell*. Conference: International Conference on Solar Energy & Photovoltaics (2012)
13. Y.M. Evtushenko, S.V. Romashkin, N.S. Trofimov, T.K. Chekhlova, Optical properties of TiO2 thin films. Phys. Procedia **73**, 100–107 (2015)
14. G. Deutscher, K.A. Müller, Phys. Rev. Lett. **59**, 1745 (1987)
15. T. Sakudo, H. Unoki, Dielectric properties of SrTiO3 at low temperatures. Phys. Rev. Lett **26**(18), 1147–1147 (1971)
16. M.N. Kamalasanan, N.D. Kumar, S. Chandra, Dielectric and ferroelectric properties of BaTiO3 thin films grown by the sol-gel process. J. Appl. Phys. **74**, 5679 (1993)
17. H. Han, C. Voisin, S. Guillemet-Fritsch, P. Dufour, C. Tenailleau, C. Turner, J.-C. Nino, Origin of colossal permittivity in BaTiO3 via broadband dielectric spectroscopy. J. Appl. Phys **113**(2), 024102 (2013). https://doi.org/10.1063/1.4774099
18. R. Legros, R. Metz, A. Rousset, Structural properties of nickel manganite NiMn3-xO4 with 0.5 <x <1. J. Mater. Sci. **25**, 4410–4414 (1990)
19. I. Terasaki, Y. Sasago, K. Uchinokura, Large thermoelectric power in NaxCo2O4 single crystals. Phys. Rev. B **56**, 12685–12687 (1997)
20. F. Chen, L. Stokes, R. Funahashi, Appl. Phys. Lett. **81**, 1459 (2002)
21. D.G. Cahill, H.E. Fischer, T. Klitsner, E.T. Swartz, R.O. Pohl, Thermal conductivity of thin films: measurements and understanding. J. Vac. Sci. Technol. A **7**, 1259 (1989)
22. J. Loureiro, J.R. Santos, A. Nogueira, F. Wyczisk, L. Divay, S. Reparaz, F. Alzina, C.M. Sotomayor Torres, J. Cuffe, F. Montemor, R. Martins, I. Ferreira, Nanostructured p-type Cr/V2O5 thin films with boosted thermoelectric properties. J. Mater. Chem. A **2**, 6456 (2014)
23. R. Venkatasubramanian, E. Siivola, B. O'Quinn, K. Coonley, T. Colpitts, P. Addepalli, M. Napier, M. Mantini, Nanostructured superlattice thin-film thermoelectric devices, in *Nanotechnology and the Environment*, (American Chemical Society, Washington, 2004), pp. 347–352
24. K. Koumoto, Y. Wang, R. Zhang, A. Kosuga, R. Funahashi, Oxide thermoelectric materials: a nanostructuring approach. Annu. Rev. Mater. Res. **40**, 363–394 (2010)
25. N.F. Mott, Conduction in non-crystalline materials. Philos. Mag. **19**(160), 835–852 (1969)
26. B. Giuot, R. Legros, R. Metz, A. Rousset, Electrical conductivity of copper and nickel manganites in relation with the simultaneous presence of Mn3+ and Mn4+ ions on octahedral sites of the spinel structure. Solid State Ionics **51**, 7–9 (1992)
27. I. Sinnarasa, Y. Thimont, L. Presmanes, A. Barnabé, P. Tailhades, Thermoelectric and transport properties of Delafossite CuCrO2: Mg thin films prepared by RF magnetron sputtering. Nanomaterials **7**, 157 (2017). https://doi.org/10.3390/nano7070157
28. A.M. Nardes, M. Kemerink, R.A.J. Janssen, Anisotropic hopping conduction in spin-coated PEDOTT: PSS thin films. Phys. Rev. B **76**, 085208 (2007). https://doi.org/10.1103/PhysRevB.76.085208
29. Z.M. Gibbs, H.S. Kim, H. Wang, G. Jeffrey Snyder, Appl. Phys. Lett. **106**, 022112 (2015)
30. W. Kobayashi, I. Terasaki, M. Mikami, R. Funahashi, Negative thermoelectric power induced by positive carriers in CaMn3-xCuxMn4O12. J. Phys. Soc. Jpn. **73**(3), 523–525 (2004)

31. T. Tanaka, S. Nakamura, S. Lida, Observation of distinct metallic conductivity in NaCo2O4. Jpn. J. Appl. Phys. **33**, L581–L582 (1994)
32. T.D. Sparks, Dissertation, Harvard University Cambridge, Massachusetts, 2012
33. S. Lee, K. Hippalgaonkar, F. Yang, J. Hong, C. Ko, J. Suh, K. Liu, K. Wang, J.-J. Urban, X. Zhang, C. Dames, S.-A. Hartnoll, O. Delaire, J. Wu, Anomalously low electronic thermal conductivity in metallic vanadium dioxide. Science **355**(6323), 371–374 (2017)
34. J.A. Thornton, High rate thick film growth. Annu. Rev. Mater. Sci. **7**(1), 239–260 (1977)
35. G. Chen, M.S. Dresselhaus, G. Dresselhaus, J.-P. Fleurial, T. Caillat, Recent developments in thermoelectric materials. Int. Mater. Rev. **48**(1), 45 (2003)
36. T.M.C. Dinh, A. Barnabe, M.A. Bui, C. Josse, T. Hungria, C. Bonningue, L. Presmanes, P. Tailhades, FIB plan view lift-out sample preparation for TEM characterization of periodic nanostructures obtained by spinodal decomposition in Co1.7Fe1.3O4 thin films. CrystEngComm **20**, 6146 (2018). https://doi.org/10.1039/c8ce01186a
37. P. Fan, Z.H. Zheng, Z.K. Cai, T.B. Chen, P.J. Liu, X.M. Cai, D.P. Zhang, G.X. Liang, J.T. Luo, The high performance of a thin film thermoelectric generator with heat flow running parallel to film surface. Appl. Phys. Lett. **102**, 033904 (2013)
38. D. Zappa, S. Dalola, G. Faglia, E. Comini, M. Ferroni, C. Soldano, V. Ferrari, G. Sbervegleri, Intergration of ZnO and CuO nanowires into a thermoelectric module. Beilstein J. Nanotechnol. **5**, 927–936 (2014)
39. A. Pérez-Rivero, M. Cabero, M. Varela, R. Ramírez-Jiménez, F.J. Mompean, J. Santamaría, J.L. Martínez, C. Prieto, Thermoelectric functionality of Ca3Co4O9 epitaxial thin films on yttria-stabilized zirconia crystalline substrate. J. Alloys Compd. **710**, 151–158 (2017)
40. I. Sinnarasa, Y. Thimont, L. Presmanes, A. Barnabe, P. Tailhades, Determination of modified ZT validity for thermoelectric thin films with heat transfer model: case of CuCrO2:Mg deposited on fused silica. J. Appl. Phys **124**, 165306 (2018)
41. F.M. Smits, Measurement of sheet resistivities with the four-point probe. Bell Syst. Tech. J. **37**(3), 711–718 (1958)
42. J.H. Bahk, T. Favaloro, A. Shakouri, Thin film thermoelectric characterization techniques, in *Annual Review of Heat Transfer*, ed. by G. Chen et al., vol. 16, (Begell House Inc., New York, 2013)
43. C.T. Prewitt, R.D. Shannon, D.B. Rogers, Chemistry of noble metal oxides: II. Crystal structures of PtCoO2, PdCoO2, CuFeO2 and AgFeO2. Inorg. Chem **10**, 719–723 (1971)
44. T. Nozaki, K. Hayashi, T. Kajitani, Electronic structure and thermoelectric properties of the Delafossite type oxides CuFe1-xNixO2. J. Electron. Mater. **38**(7), 1282–1286 (2009)
45. A.P. Amrute, G.O. Larrazabal, C. Mondelli, J. Pérez-Ramirez, CuCrO2, Delafossite: a stable copper catalyst for chlorine. Angew. Chem. Int. Ed. **52**(37), 9772–9775 (2013). https://doi.org/10.1002/anie.201304254
46. C. Ruttanapun, S. Maensiri, Effects of spin entropy and lattice strain from mixed-trivalent Fe3+/Cr3+ on the electronic, thermoelectric and optical properties of delafossite CuFe1−xCrxO2 (x = 0.25, 0.5, 0.75). J. Phys. D. Appl. Phys **48**, 495103 (2015). https://doi.org/10.1088/0022-3727/48/49/495103
47. E. Guilmeau, M. Poienar, S. Kremer, S. Marinel, S. Hébert, R. Frésard, A. Maignan, Mg substitution in CuCrO2 delafossite compound. Solid State Commun. **151**(23), 1798–1801 (2011)
48. W. Koshibae, K. Tsutsui, S. Maekawa, Thermopower in cobalt oxides. Phys. Rev. B **62**(11), 6869–6872 (2000)
49. A. Maignan, C. Martin, R. Frésard, V. Eyert, E. Guilmeau, S. Hébert, M. Poienar, D. Pelloquin, On the strong impact of doping in the triangular antiferromagnet CuCrO2. Solid State Commun. **149**, 962–967 (2009)
50. A.J. Bosman, H.J. Van Daal, Small polaron versus band conduction in some transition-metal oxides. Adv. Phys. **19**(77), 1–117 (1970)
51. S. Saini, P. Mele, K. Miyazaki, A. Tiwari, On-chip thermoelectric module comprised of oxide thin film legs. Energy Convers. Manag. **114**, 251–257 (2016)

Editorial Note

Dear Readers,

I am pleased to deliver the Special Book "Thermoelectric thin films: materials and devices", inspired from the Symposium A-5 "Thermoelectric materials for sustainable development" during conference IUMRS-ICAM 2017 in Kyoto University.

The authors of this book are distinguished colleagues and friends, most of them participants and contributors to symposium A-5 in IUMRS-ICAM 2017. After almost one-and-a-half year, this book was finally edited and delivered today.

On behalf of all the editors, I would like to warmly thank our colleagues for their wonderful contributions, and everyone who contributed with their precious help during the revision and editing of this book.

Tokyo, Japan   Paolo Mele (Shibaura Institute of Technology, Japan)
March 11, 2019

© Springer Nature Switzerland AG 2019   205
P. Mele et al. (eds.), *Thermoelectric Thin Films*,
https://doi.org/10.1007/978-3-030-20043-5

Index

© Springer Nature Switzerland AG 2019
P. Mele et al. (eds.), *Thermoelectric Thin Films*,
https://doi.org/10.1007/978-3-030-20043-5

Printed in the United States
By Bookmasters